D1767740

Sensor Technology in Neuroscience

RSC Detection Science Series

Editor-in-Chief:
Professor Mike Thompson, *University of Toronto, Canada*

Series Editors:
Dr Subrayal M Reddy, *University of Surrey, Guildford, UK*
Dr Damien Arrigan, *Tyndall National Institute, Cork, Ireland*

Titles in the Series:
1: Sensor Technology in Neuroscience

How to obtain future titles on publication:
A standing order plan is available for this series. A standing order will bring delivery of each new volume immediately on publication.

For further information please contact:
Book Sales Department, Royal Society of Chemistry, Thomas Graham House, Science Park, Milton Road, Cambridge, CB4 0WF, UK
Telephone: +44 (0)1223 420066, Fax: +44 (0)1223 420247
Email: booksales@rsc.org
Visit our website at www.rsc.org/books

Sensor Technology in Neuroscience

Michael Thompson, Larisa-Emilia Cheran
University of Toronto, Canada
Email: mikethom@chem.utoronto.ca

and

Saman Sadeghi
University of California Los Angeles, USA

RSCPublishing

QH
581.2
.T46
2013

RSC Detection Science Series No. 1

ISBN: 978-1-84973-379-3
ISSN: 2052-3068

A catalogue record for this book is available from the British Library

© M Thompson, L-E Cheran and S Sadeghi 2013

All rights reserved

Apart from fair dealing for the purposes of research for non-commercial purposes or for private study, criticism or review, as permitted under the Copyright, Designs and Patents Act 1988 and the Copyright and Related Rights Regulations 2003, this publication may not be reproduced, stored or transmitted, in any form or by any means, without the prior permission in writing of The Royal Society of Chemistry or the copyright owner, or in the case of reproduction in accordance with the terms of licences issued by the Copyright Licensing Agency in the UK, or in accordance with the terms of the licences issued by the appropriate Reproduction Rights Organization outside the UK. Enquiries concerning reproduction outside the terms stated here should be sent to The Royal Society of Chemistry at the address printed on this page.

The RSC is not responsible for individual opinions expressed in this work.

Published by The Royal Society of Chemistry,
Thomas Graham House, Science Park, Milton Road,
Cambridge CB4 0WF, UK

Registered Charity Number 207890

For further information see our web site at www.rsc.org

Foreword

Sir John Eccles, along with Andrew Huxley and Alan Lloyd Hodgkin, received the 1963 Nobel Prize in physiology for their work on understanding synaptic communication between neurons. Elucidation of the synapse was essentially a paradigm shift in understanding how the brain functioned. Yet, despite this achievement, Eccles recognized that their discovery was but a stepping stone. Just before his death in 1997, Eccles wrote in his final book *How the Brain Controls Itself* (1994), *'The more we discover scientifically about the brain the more clearly do we distinguish between the brain events and the mental phenomena and the more wonderful do the mental phenomena become'*. He recognized that a reductionist approach was required to understand the mechanisms of the brain. But with each discovery, the complexity of the brain, thought processes and consciousness unveiled themselves layer by layer, leading each time to a new understanding of the brain, which in itself, was more exciting and mysterious than the previous understanding.

Since the time of Eccles' discoveries, we have witnessed the development of numerous advances in neuroscience that include the delineation of molecular and ionic regulation of memory, neurogenesis, the molecular biology revolution, synaptic organization of brain nuclei, discovery of novel neurotransmitters and neuromodulators, and genomic sequencing. Yet despite this, at the time of writing in 2013, the study of the brain is still in its infancy. Discovery is limited by the available tools. The microscope and its various forms, and cellular imaging methods are key examples. But the development of such tools and the interpretation of the resulting observations are dependent upon asking the appropriate questions. History has taught us that sometimes the best questions are the result of explaining the observations of one discipline with the theories of another.

The quest to develop a set of tools and technologies that can bridge the various disciplines of neuroscience from psychology to neurology remains a key challenge. For example, the *in vitro* culture of neurons is one such tool that has enabled neuroscientists to study and manipulate these cells in a manner that was previously thought impossible. With this technology, neurons can be readily grown in petri dishes and easily studied. But can we relate the physiology and behaviour of a cell in a dish to that in the brain? No, not in all respects because, despite the complex biology of a neuron, cell-to-cell and cell-to-substrate interactions are ultimately physicochemical in nature. Altering the association of cells with each other or to their substrates will affect their morphology, membrane function and signalling capacity. We have to remember that neurons have been evolving for perhaps close to a billion years and the basic cell longer than three billion years. Over this period, the optimal activity of neurons is intrinsically linked to their intimate co-evolution with such proximal factors as their cellular neighbours and substrate proteins in the form of the extracellular matrix. Thus, introducing neurons into the artificial environment associated with our current *in vitro* culturing technology will alter the nature of physicochemical interactions of cell interfaces with their ambient environment leading to an altered function of the neurons we wish to study. In order to understand the relationship between *in vitro* and *in vivo* brain cell function, we require methods to understand how this physicochemical environment differs and how it relates to cell function.

We, as neuroscientists, understand this. To avoid this problem, we utilize a variety of different techniques to examine morphology, metabolism, signalling, proliferation and gene transcription, for example, to provide an integrated view as to how these cells react *in vitro* and how these findings relate to similar cells in the brain. This is a complex and laborious process that requires a comparison of the system under study with similar investigations performed under various *in vitro* and *in vivo* conditions. However, if it was possible to examine the performance of these cells utilizing a system than incorporated the integrated physiology of cell function, rather than the separate expression of each of numerous biomarkers, then the task of interpretation would be considerably easier. Such data could provide a stronger rationale for designing *in vitro* experiments that would reflect the *in vivo* situation.

The technological and theoretical approaches described in this book consider this problem. Utilizing novel sensor technology, these investigations have the potential to relate the overall cell function *in vitro* to *in vivo* performance. For example, we understand that synchronous neuronal activity in the brain is an essential component of normal neurological processes. But how do we know if the brains cells *in vitro* display a similar pattern of synchronous behaviour to that observed *in vivo*? How can we relate the neurotransmitters such as glutamate, gamma-aminobutyric acid (GABA) and acetylcholine, for example, with the associated ion flux in neurons associated with sodium, potassium, chloride or calcium when examining differing *in vitro* and *in vivo* model systems? A comparison of these processes *in vitro* to *in vivo* situations is essential to understand the basic elements of neurophysiology. The technologies described in this book consider these issues and provide a new approach to circumventing these problems. More importantly, however, are the opportunities for the major discovery of neural processes by investigating physical characteristics that have largely been ignored in neuroscience.

On a practical level, such methods can be used to determine the relevance of particular *in vitro* methods to what is observed *in vivo*. With modern techniques of nuclear magnetic resonance imaging and positron emission tomography, which utilize physicochemical data as opposed to biological interactions, we can get a sense of how the brain is integrated in the interpretation of certain stimuli. However, because of our current limited understanding of neural functions, this renders an imperfect picture and requires integration of our studies in the laboratory with clinical studies.

An integrated understanding of the brain is essential to understand the complex interaction of thought processes, cognition, consciousness or, indeed, understanding how psychogenic drugs affect these processes. Until this is achieved, we will not be able to understand such neuropsychiatric conditions as depression and anxiety, neurodegeneration, epilepsy, aging or autism, to name a few. Recent studies of electrical stimulation, transcranial magnetic stimulation and deep brain stimulation have been successful in treating a number of neurological impairments, yet we only have a partial understanding of their mechanisms. In 1948, Bertrand Russell, in his book, *Human Knowledge, Its Scope and Limits*, argued that there was a fundamental separation between the science of psychology and the science of physics in understanding the brain. I would posit that this concern is still true of today. Technologies and new neurological theory will be required to bridge the gaps.

Undoubtedly, the study of neuroscience still has a long way to go. However, I expect that the exciting new technologies, tools and approaches described within the chapters of this book will have a significant impact on the study of neuroscience in the near future.

David A. Lovejoy
Department of Cell and Systems Biology
University of Toronto
Author of *"Neuroendocrinology, An Integrated Approach"*

Preface

The authors of this text have a high level of interdisciplinary expertise in terms of biosensor design, biophysical chemistry, surface science, materials science, molecular imaging and microelectronic technology. In this regard, it has become evident in recent times that neuroscience is an incredibly wide-ranging field with research contributions emanating from many areas of science, medicine and engineering. The focus of this text is on potential devices and instruments that are currently in use, or being developed, to aid biosensor specialists and neuroscientists alike in their explorations.

The text is modular in character. We anticipate a readership that originates from a disparate set of research communities. The reader might wish to zero in on sections where there is less familiarity. For example, the work opens with a concise look at the world of biosensor technology and deals with the design and physics of these devices and how they are employed in the detection of biochemical events. Biosensor specialists may well 'skip' this chapter and proceed on to other aspects of the text. However, this type of analytical device may be of interest from a learning perspective to the more biologically inclined reader. The second chapter discusses the interaction of cells of various types with surfaces, certainly a key aspect of any treatment of signalling neurochemistry and brain functions in general. This includes a discussion of work on surface functional group chemistry and the more morphological aspects of surface–cell interactions. Many topics here will already be well-known to the cell biologist and the neuroscientist, but analytical chemists and biosensor experts may find this material useful. Next, we deal with the one of the most rapidly developing areas of endeavour in neuroscience, that of the design and implementation of electronic devices for the detection of neurobiochemical and other events. Chapter 4 presents a natural extension of the previous section in terms of a concise journey into the employment of 'nano' based devices into brain research. We then proceed to a foray into a special group of techniques for the study of neurons, and other cells, based on probes that are subjected to vibration. This includes impedance spectroscopy and the two label-free biosensor methodologies, surface plasmon resonance and acoustic wave physics. No book on sensors used in neuroscience would be complete without examining implantable devices which are in direct contact with neural tissue (Chapter 6). Finally, your authors have a go at gazing into the proverbial crystal ball in Chapter 7.

Given their passion for interdisciplinary science, the preparation of this tract has been a labour of love for your writers. In this sprit we fervently hope that the book will be of interest to those individuals engaged in research from such wide-ranging fields such as analytical chemistry, engineering and neurology.

Finally, we would like to express our deep appreciation to all past and present members of the Thompson Biosensors Group at the University of Toronto for their much valued research contributions and unfailing support. We also wish to thank Eddie Chan and Sean Robertson of the University of Toronto for their assistance in preparing the book chapters.

Michael Thompson
Larisa-Emilia Cheran
Saman Sadeghi

'The paradox of neuroscience is that its astonishing progress has exposed the limitations of its paradigm, as reductionism has failed to solve our emergent mind'.

'If neuroscience is ever going to discover the neural correlates of consciousness, or find the source of the self, or locate the cells of subjectivity—if its ever going to get beyond a glossary of our cortical parts—then it has to develop a new method, one that's able to construct complex representations of the mind that aren't built from the bottom up'.

Jonah Lehrer, Why science needs art,
Shift: At the Frontiers of Consciousness,
No. 20, 2008, Institute of Noetic Sciences,
Petaluma, California, USA, p. 10

Contents

Chapter 1	Introduction to Biosensor Technology	1

    1.1   Sensor Anatomy, Signaling and Properties   1
    1.2   Genesis of Biosensor Technology   4
    1.3   Probe Attachment to the Sensor Structure   5
        1.3.1   Direct and Linker Adsorption   6
        1.3.2   Entrapment and Encapsulation   7
        1.3.3   Covalent binding   8
        1.3.4   Assembled Monolayer Chemistry   10
        1.3.5   The Molecularly Imprinted Polymer   12
    1.4   Device Transduction of Biochemical Interactions   13
        1.4.1   Electrochemical Systems   13
        1.4.2   Acoustic Wave Physics and Devices   19
        1.4.3   Electromagnetic Radiation: Optical Devices   28
        1.4.4   Brief Summary of the Adjunct Technology Approach   39
    References   41
    Selected Bibliography of Biosensor Technology 1987–2012   46

Chapter 2	The Cell-Substrate Surface Interaction	50

    2.1   Cells and Surfaces   50
    2.2   Substrate Surface Parameters: A Précis   51
    2.3   The Eukaryotic Cell and Environment: A Précis   53
    2.4   The Neuron: A Précis   54
        2.4.1   Anatomy and Types   55
        2.4.2   Action Potential and Electrical Conduction   58

RSC Detection Science Series No. 1
Sensor Technology in Neuroscience
By Michael Thompson, Larisa Emilia Cheran and Saman Sadeghi
© M Thompson, L-E Cheran and S Sadeghi 2013
Published by the Royal Society of Chemistry, www.rsc.org

	2.5	Cell Adhesion, Growth, Guidance and Proliferation on Substrates		60
		2.5.1	General Considerations	60
		2.5.2	Bare Substrates	61
		2.5.3	Polypeptide Coating	64
		2.5.4	Extracellular Matrix Proteins and Derived Peptides	65
		2.5.5	Substrate Morphology	72
		2.5.6	Substrate Rigidity and Elasticity	77
	2.6	Biocompatibility and the Substrate–Blood and Platelet Interaction: A Comment on Long-term Effects		80
	References			82
Chapter 3	Electronic Detection Techniques			87
	3.1	A Review of Neuron Field Potentials		87
	3.2	Cultured Neurons and Neuro-electronic Interface		88
	3.3	Charge Transfer and the Interface		93
	3.4	Field Effect Transistors as Neurotransducers		95
	3.5	Microelectrode Array Structures		97
	3.6	Microelectronic Interfaces for *In Vitro* Study of Neurons		102
	3.7	Fabrication		105
		3.7.1	Regeneration Sieves and Cone-ingrowth Electrodes	106
		3.7.2	Microfluidic Structures	107
		3.7.3	Self-assembled Networks	109
	3.8	*In Vitro* Microelectrodes Arrays		110
	3.9	Microfluidics in Neurobiological Research		112
	3.10	Biosensors for Neuroscience Applications		115
		3.10.1	*In Vitro* Microelectronic Interfaces	115
		3.10.2	Microscale Cell Culture Analogues	116
		3.10.3	Microelectrode Arrays in Drug Discovery	116
		3.10.4	Microelectrode Arrays in Toxicology	118
		3.10.5	Microelectrode Arrays in Basic Neuroscience Research	122
	References			126
Chapter 4	Nanosensing the Brain			130
	4.1	Nanoparticles as Reporters of Brain Activity		131
	4.2	Nanotubes and Nanowires		132
	4.3	Graphene		135

	4.4	Applications of Nanotechnologies in Neuroscience	137
		4.4.1 Nanostructures as Scaffolds for Neuroregeneration and as Interface for Sensing and Stimulation	137
		4.4.2 Nanoribbons for Sensing Cellular Deformation	139
	4.5	Challenges and Future Perspective of Nanotechnologies in Neuroscience	139
	References		140

Chapter 5 The Vibrational Field and Detection of Neuron Behavior — 142

5.1	Extending Human Sensory Capabilities	142
5.2	The Vibrational Field as a Neural Sensor Platform	143
	5.2.1 The Simple Vibrating Probe	144
	5.2.2 Electric Impedance Sensing of the Cell–Substrate Interaction	146
	5.2.3 Miniaturization of the Electrical Impedance Tomography Technique	146
	5.2.4 Optical Sensing Platforms	154
	5.2.5 Acoustic Wave Detection	157
	5.2.6 Origin of Oscillations and Neuronal Resonance	160
	5.2.7 The Scanning Kelvin Nanoprobe	165
5.3	Future Possibilities in Cellular and Neuronal Detection	169
References		170

Chapter 6 The Biomimetic Interface between Brain and Electrodes: Examples in the Design of Neural Prostheses — 172

6.1	The Nature of the Device–Brain Interface	172
6.2	Electrode–Tissue Interface in Deep Brain Stimulation	175
	6.2.1 Electrode Implantation	175
	6.2.2 Device-related Complications	179
6.3	Motor Cortex Prostheses	180
6.4	Replacing Damaged Brain Components	183
6.5	Retinal Prosthetic Interfaces	186
6.6	Technological Advances and Novel Strategies for Improving the Electrode–Brain Interface	189
References		191

Chapter 7 A Look at the Future — 194

| 7.1 | Quantum Neurobiology | 194 |
| 7.2 | Nanoneuromedicine | 195 |

7.3	Neuropharmacology	196
7.4	Genetics	196
7.5	Cognitive Enhancers	197
7.6	Brain Imaging	198
7.7	Stem and Cancer Cells	198
7.8	Regenerative Techniques in Neuroscience	199
7.9	Prosthetic Implants	199
7.10	Dementia, Alzheimer's Disease and Reversing the Ageing Process in the Brain	200
7.11	The Human Brain Project and Computer Simulation	201
7.12	Conservative Perspectives: A Final Comment	201

Subject Index   **203**

CHAPTER 1
Introduction to Biosensor Technology

1.1 Sensor Anatomy, Signaling and Properties

The notion of a sensor device is common knowledge to all. The range of these structures in modern times is immense, ranging from simple physical measurements such as temperature to complex devices that incorporate human cells in their design. The number of applications is also numerous including industrial processing, pharmaceutical analysis, automotive operation, military technology and environmental signaling to name just a few areas of use. In this section we introduce the basics of a special branch of sensor technology that deals with the detection of chemicals, with relevance to the research in neuroscience described later in the chapter. The emphasis is on devices, which constitute the main structures employed in biosensor technology, rather than a comprehensive review of the field.

Devices that detect and signal the presence of chemicals have evolved through two different pathways, although the distinction between the two is somewhat arbitrary. The structure is composed of a chemical recognition site attached to a substrate surface which, in turn, is in close proximity and union with a transducer. Such a system can be used to respond selectively to the presence of chemicals we term the target or analyte, either in the gas or liquid phase. The chemical recognition site is often referred to as the receptor or probe. The technology relies on the ability of such a configuration to 'recognize' chemicals through selective binding at the substrate surface of the device with such surface presence being converted into an electrical signal *via* the particular physics of a transducer. Chemical sensors are generally considered to involve non-biological/biochemical probes in their design and the same is often regarded to be the case for the analyte. A simple example is the well-known tin

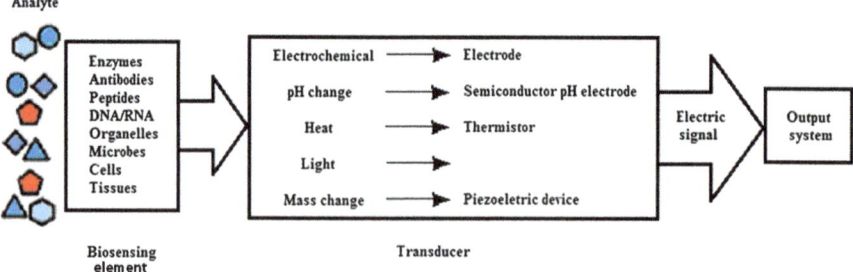

Figure 1.1 Schematic of typical components of biosensor architecture.

oxide sensor which responds to gases such as carbon monoxide.[1,2] In contrast, biosensors are composed of a union between a transducer and biological/biochemical receptor. A schematic of the structure including transduction types is depicted in Figure 1.1. Note that the probe can consist of a variety of biological entities ranging from antibodies to live biological cells Given that nucleic acids and proteins are chemicals, as are the targets that cells as probes are designed to detect, it is clear that the distinction between chemical- and biosensors is artificial.

Crucial aspects of biosensor technology are the nature of the response of the device with respect to time and whether ancillary chemistry is required in addition to the basic probe in order to achieve a signal. With respect to the former point, there are sensor specialists who take the view that such a device must respond to its analyte in real time. Conversely, the field is often considered to include 'one-test' disposable structures where there is no attempt to conduct a measurement over a period of time, except in a repeated dipstick fashion. The ubiquitous pregnancy and glucose test strips that are widely commercially available constitute examples of this approach to biodetection. The use of an adjunct chemical in addition to the receptor to achieve a transducer signal is often termed tagging or labeling. An example of this strategy is the use of dyes in conjunction with nucleic acid probes in order to produce fluorescent signals.[3] In certain cases, there is insufficient intrinsic fluorescence in nucleic acid molecular probes to allow the direct possibility of detection. The same is true for electrochemical methods where organometallic complexes (of Ru) have to be employed for work with nucleic acids in order to detect redox chemistry. Technology where such an approach is avoided is called 'label-free detection' and is often regarded to be attractive in view of the fact that sensor fabrication becomes a somewhat simpler process.

Additional important technical factors are the possibilities for incorporation of the device in flow-through automation, sensor miniaturization and prevention of non-specific adsorption. Such automation involving standalone systems avoids time-consuming personal intervention and allows rapid data collection and validation. Microfluidic systems offer speed and saving of reagent costs. Non-specific adsorption of unwanted components on the device surface poses something of an Achilles' heel for biosensor technology. The selectivity and limit of detection of the sensor when used, for example, in blood, serum, urine and tissue will clearly be influenced strongly by interfering components of the biological sample. It appears to be the case that wide scale use of biosensors in, for example, clinical biochemistry, has not occurred primarily because of this issue.

The placement of a solid transducer–probe combination into a biological sample will result in a signal originating from a composite response, X_n, according to the following matrix:[4]

$$X_n = S_{n1}C_1 + S_{n2}C_2 + \cdots S_{nn}C_n \quad (1.1)$$

where C and S values represent the concentration of analyte and interferants in proximity to the device, and the response sensitivities, respectively.

An enormous number of components would be expected to be involved in this equation in terms of biological samples. A sensitive response (*e.g.* volts or amps) implies maximization of one S value and, for a selective signal, a minimization of all other S values. To a first approximation, it is necessary to trap the analyte on the device surface to allow the sensor response and, as mentioned above, to repel or avoid such binding of all other elements. Accordingly the sensor signal will be a composite of the chemistry of the attachment process and the physical perturbation caused by the probe–analyte complex. This leads to some interesting aspects concerning the nature of the couple between physical chemistry and the transduction process. In certain cases, as will be seen in later sections, the mere presence of the analyte can influence the transducer and the resulting signal is often referred to as a 'mass response'. However, a situation can be envisaged where a structural shift, such as a probe conformational change, and regardless of whether the response is related to final state effects or the change itself, is required for detection to take place. The physical chemistry of transduction in this case is reminiscent of agonist *versus* antagonist interactions and will be familiar to the biochemist community. All these mechanisms will obviously be an intrinsic component of the sensitivity parameter, S_{nn}, outlined above.

In summary, there are a number of key desirable properties that a biosensor should possess, although some features are of course more important for some applications than others:

- Selectivity or even specificity (see ref. 4 for an excellent definition of these parameters) with respect to the response to the analyte, as described above, is a given. Non-specific binding through adsorption to the device surface by interferants will clearly have a strong influence on selectivity, especially if such components are at a high concentration .in the sample under analysis.
- High sensitivity is required with a resulting low value of limit of detection.
- High accuracy is required in terms of concentration measurement for the analyte. Signaling must be conducted with high reproducibility—a precision issue.

- The dynamic range with respect to response should be as high as possible for certain targets (for example, clinical values of an analyte may vary considerably).
- With regard to real-time detection, speed of response is crucial. For the one-time determination device this aspect may not be so important.
- Device response calibration in terms of concentration is mandatory and especially important for the real-time sensor. This feature has constituted an extremely difficult problem when it comes to, for example, the operation of a corporeal implantable structure. Early solutions to this issue may well lie in a strategy of device self referencing.[5]
- User ease of use may well be a factor especially where expertise in device understanding is lacking.
- There are ancillary issues such as robustness and cost which may be crucial from a more commercial point of view.

A bibliography of texts that deal with biosensor technology and related subject matter is provided at the end of this chapter.

1.2 Genesis of Biosensor Technology

The advent of biosensor technology has been widely attributed to the work of Clark and Lyons,[6] which involved an electrochemical amperometric mechanism for the measurement of glucose concentrations in biological samples. Prior to this system (1950s), Clark *et al.*[7] were working on the electrochemical detection of oxygen dissolved in blood and tissue *via* O_2 reduction, and this research served as something of a precursor to the subsequent developments with glucose.

The basis of the original glucose measurement was an enzyme electrode (Figure 1.2) in which a layer containing the enzyme, glucose oxidase (GO) was

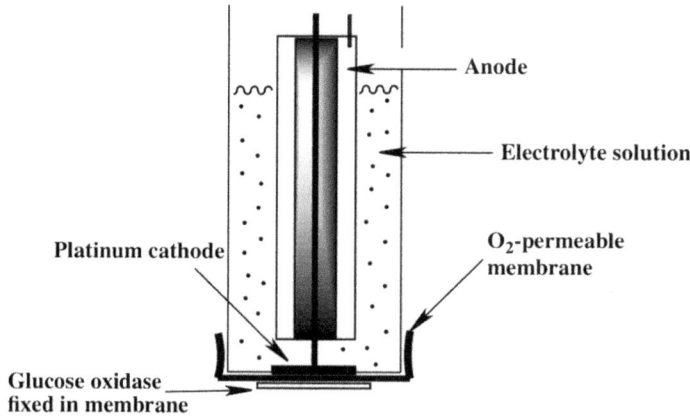

Figure 1.2 Schematic of a typical design of an enzyme (glucose oxidase) electrode.

placed in close proximity to a platinum electrode polarized at a potential of +0.6 V. The substrate reacts with the enzyme according to:

$$\text{glucose} + O_2 \xrightarrow{\text{GO}} \text{gluconic acid} + H_2O_2$$

The Pt electrode is designed to respond to the hydrogen peroxide produced by the enzyme–glucose reaction. This progress led to a large number of O_2 mediated devices being developed which included a variety of enzymes and substrates. A further early example connected to glucose detection was the work of Updike and Dicks[8] who immobilized a gel impregnated with the oxidase on a semi-permeable membrane. In this experiment, the amperometric signal resulted from changes in the partial pressure of oxygen as it diffused through the membrane to a sensor. Not surprisingly other components of the basic oxidase reaction have been studied such as pH changes associated with the production of gluconic acid; in this situation the electrochemistry involved potentiometry.

Although a large number of enzyme-based systems have been developed over many years, it is fair to say that by far the most research has centered on glucose because of the obvious significance to the diabetic patient. Literally hundreds if not thousands of scientific papers have been devoted to this substrate. In modern times, O_2 as a redox mediator is avoided and other species capable of electron transfer are employed to serve the same purpose.[9,10] Indeed such a strategy is very much a component of the commercial test strips that are so familiar to individuals suffering from diabetes.

We now turn to a concise look at contemporary biosensor technology using a device outside-to-inside approach.

1.3 Probe Attachment to the Sensor Structure

In order for a biosensor to function it is mandatory to couple the probe or receptor to the solid surface of the particular transducer being employed, and if not bound at the device interface, it must generally be arranged be in close proximity. Over the years a plethora of methods have been developed for this purpose,[11,12] and although some of these are common to various probes, a large number have been designed for a specific receptor type. The overall goal of much of this chemistry is to achieve the maximum signal with respect to operation of the device in use. In this respect there are a number of key factors at play.

- It is obviously critical to retain the binding activity of the probe for its target. It is anticipated that certain proteins, for example, may be partially denatured on attachment to the sensor surface. Others such as antibodies may be more robust. Such is generally the case for DNA and oligonucleotides whereas RNA moieties are considered to be more prone to alteration in structure. When it comes to cells it is clearly important for

them to be active and alive; this area is discussed in detail in the following chapter.
- The particular orientation of the probe with respect to the plane of the device surface is an important element. As an example, if the Fab region of an antibody is 'hidden' from the target because of binding of this region to the solid surface instead of the Fc component, a true immunochemical interaction cannot be expected.
- In terms of sensitivity it is crucial to maximize the density of probe molecules per surface unit area with the caveat specified in the next point. Such density may well be influenced by the roughness and surface area of the device. Furthermore, there are cases where polymer films are employed rather than plane surfaces but the same criterion holds in terms of overall surface density. In this situation a key factor is the ability of the analyte to diffuse into the film in order to bind to probes attached to polymer strands.
- The spatial characteristics of the probe with regard to the surface plane are a factor that is important but not studied widely. If probes are in close molecular proximity on the surface this may result in steric hindrance to target binding. Another consideration is the possibility that the chemistry of probe attachment may result in 'island' formation. In this case the overall number of binding sites will be significantly reduced. As would be anticipated, control of this chemistry is very difficult to achieve.
- There are other more mundane factors such as the cost of reagents and the potential longevity of the modified sensor surface. These are clearly critical from a commercial standpoint.

We take a concise look at the myriad of possible strategies for probe attachment to devices. The particular surface modifications necessary to impose cells on a device is considered in the following chapter.

1.3.1 Direct and Linker Adsorption

This constitutes the simplest method for placing a biomolecule onto a substrate, although there are inherent dangers for the technique in terms of possible undesirable conformational changes. The probe of interest is chemically or physically adsorbed from solution directly, in most cases, on the substrate surface of the device in use. Various protocols are employed to achieve such an effect such as dip casting, painting, spraying and spin coating—these terms will be self-explanatory to the reader. The interaction of the probe with the surface is characterized primarily by hydrogen bonding, van der Waals forces and various dipole electrostatic interactions. The energy of the attachment process is of the order of 10–40 kJ mol^{-1} and is highly dynamic in nature. The latter may lead to a lack of stability in terms of longevity of sensor performance.

It has been considered that a major aspect of this type of chemistry, when it comes to protein molecules, is the hydrophobic interaction with a hydrophobic substrate. In this case protein molecules can change their conformation in order to expose more hydrophobic domains.[13] This sort of effect will clearly be deleterious if the native tertiary structure of the molecule is altered with respect to target binding. Although hydrophobic effects can be damaging as indicated, this sort of effect is by no means restricted to non-polar surfaces. Our laboratory showed conclusively that the well-known protein, avidin, is compromised with regard to its structure by hydrophilic interfaces. Presumably, the reason for this lies in the instigated interactions of surface polar functional groups with the carbohydrate moieties of the protein molecule.[14]

The last comment leads to a look at the use of intervening adsorbed molecules as a linker for eventual biomolecule attachment. A system which is employed extremely widely is avidin/streptavidin/neutravidin chemistry.[15–17] Avidin is a tetrameric protein that contains four binding sites for the ligand, biotin. Interaction of the protein with this molecule results in a particularly strong bond ($Kd = 10^{-15}$). Accordingly, various biomolecules can be modified with relative facility by biotin addition in order to link them to surface-attached avidin, or a sister molecule such as the deglycosylated version, neutravidin. With respect to device operation, particularly where on-line detection is involved, it is common practice to introduce the protein into the system where it is allowed to simply adsorb to the device surface. Although four biotin binding sites are present, there is no doubt that at least two of these are expected to be unavailable on steric grounds for the reasons outlined above.

An analogous chemical system is the use of protein A.[18] This is a polypeptide (molecular weight 42 kD) isolated from *Staphylococcus aureus* which is capable of binding specifically to the Fc region of various antibody molecules and has thus been employed in immunosensor technology. One strategy similar to the case of avidin outlined above is to introduce this protein (for surface adsorption) in a system prior to antibody addition.

For both avidin and protein A, it is also feasible to attach the molecules to the device surface in a chemically bonded fashion which leads to a large portfolio of methods for biomolecule binding, that of using covalent bonds in attachment (see section 1.3.3).

1.3.2 Entrapment and Encapsulation

In this approach the recognition molecule is trapped within the three-dimensional structure of a specific chemical matrix. The resulting entity is placed on the device surface for transduction as described above. The matrix may simply act as a 'holding' moiety or be modified in some way to take part in the transduction process. For the former, polymers such as polyacrylamide have been employed in a number of experimental protocols. Examples are derivatization of a protein with the monomer followed by polymerization and direct diffusion of the biomolecule into a preformed polymer gel. It is also possible to attach the probe to polymer beads to be followed by cross-linking. There are a number of potential disadvantages to this type of approach for placing the probe on the device surface including the possibility of biomolecule

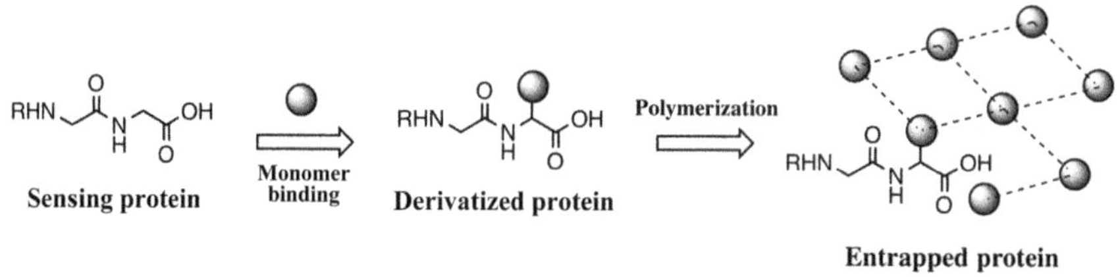

Figure 1.3 Entrapment approach to attach a protein. Biomolecule is bound to a monomer prior to polymerization.

leakage and/or denaturation. An additional consideration is the necessity for the target to diffuse into the polymer matrix in order for the biochemical interaction to take place. This process may result in a slow kinetic response in terms of transduction. Despite these problems the strategy has been employed widely in surface plasmon resonance (see later).

An analogous procedure to the one described above is the use of sol-gel technology, which not only serves to immobilize a biomolecule on a surface but also to potentially stabilize it with regard to its tertiary structure.[19] This technology has been exploited for many years with respect to applications in surface coating, production of nanomaterials and optomechanical structures. A solution of a monomer such as an alkoxide (the sol) is induced to polymerize into a biphasic configuration (the gel) which incorporates both liquid and solid. A typical monomer among many is tetraethylorthosilicate [TEOS, $Si(OC_2H_5)_4$], which is readily hydrolyzed by water to produce a siloxane bond-based polymeric structure with a gel consistency. When it comes to placement of the biomolecule, usually a protein, into the gel the term encapsulation is often used rather than entrapment. A typical approach is shown in Figure 1.3. Particularly interesting in terms of this technology is the potential of the process to stabilize the tertiary structure and conformation of the biomolecule, although it remains to be seen if this result applies to a large number of protein molecules.[20]

1.3.3 Covalent binding

With regard to the two-dimensional attachment of biomolecules to transducer surfaces in a planar format it is certainly the case that attachment *via* covalent bonds has been by far the most used approach. A very wide variety of chemistries have been employed with modest success in terms of the criteria outlined above.[21,22] Many functional groups, whether directly present on the device substrate or obtained by modification, have been utilized to form a probe partial monolayer. Examples of these are given in Table 1.1. Such groups are also available on biomolecules to instigate the surface link and examples of these for proteins are presented in Table 1.2. Oligonucleotides are very often attached to the surface through functionalized (–OH, –NH_2, –COOH, –SH, *etc.*) short alkyl chains bound to the 3′ or 5′ terminus of the nucleic acid chain. Single-strand DNA and RNA can be attached *via* a similar strategy, but the

Table 1.1 Target functional groups for biosensor immobilization of proteins.

Functional group	Amino acid
Thiol (S–H)	Cysteine
Amine (N–H, H)	Lysine (side chain)
Phenol	Tyrosine
Imidazole	Histidine
Indole	Tryptophan
Guanidine	Arginine
Carboxylic acid	Glutamic acid, aspartic acid

Table 1.2 Electrochemical techniques available for application in biosensor technology.

Technique	Methodology
Potentiometry	Measures electrochemical potential at current $=0$
Voltammetry	Measures current at applied potential; controlled concentration of electroactive species
Amperometry	Measures current versus concentration; potential is controlled
Coulometry	Measures charge over time with controlled potential
Conductometry	Measures conductance ($=1/$resistance) at controlled concentration

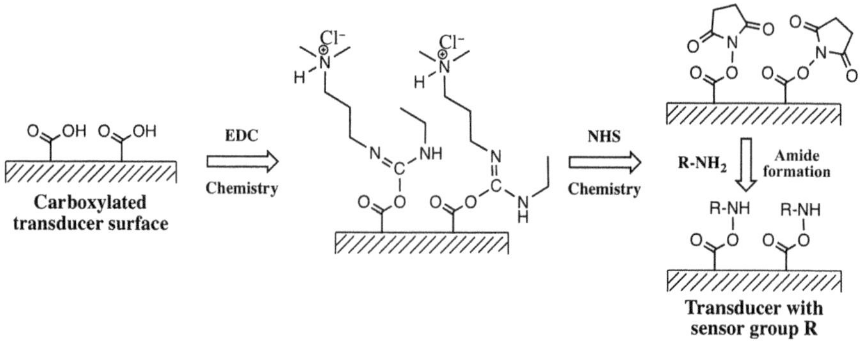

Figure 1.4 An example of covalent surface attachment of a protein molecule *via* an activation process.

base residues, especially thymine, can also be the groups of choice though, in this case, linking may compromise future duplex formation.

A common protocol to bind enzymes, antibodies and molecular receptors to the substrate is to initially functionalize the surface, followed by a second reaction(s) which is usually termed 'activation'. In essence, this process simply allows a convenient, highly reactive 'linker' moiety to be introduced to the system. The resulting interface is then normally allowed to react in turn with one of the protein nucleophilic groups mentioned above. The literature is replete with many examples of this sort of approach and Figure 1.4 provides a schematic of one strategy. If groups already evident on the device surface are not used directly, functionalization of surfaces is often achieved with species such as aminopropyltriethoxysilane (APTES) which reacts with interfacial hydroxyl groups on whatever substrate they are present. Many other chemical systems have been employed. One activating configuration that has been used ubiquitously in the literature is the well-known ethyl(dimethylaminopropyl) carbodiimide–*N*-hydroxysuccinimide (EDC-NHS) chemistry (see Figure 1.4).

1.3.4 Assembled Monolayer Chemistry

The introduction of close-packed monolayers of either membrane-forming molecules or linking systems for attachment of probes, often following similar activation to that described above, have been used widely in biosensor development. Two very different strategies are employed, the first being the Langmuir–Blodgett film technique.[23,24] In this experiment, close-packed monolayers are imposed on a surface in an orchestrated fashion by transferring lipid or lipid-like (amphiphile) films from the Langmuir trough, under the correct surface pressure conditions, by a dipping process (Figure 1.5). In principle, several layers can be deposited in sequence using the dipping approach. The important advantage offered by this technique is the possibility to combine artificial lipid membrane configurations directly and *in situ* with integral membrane proteins (IMPs). Such membrane systems are, of course,

Introduction to Biosensor Technology 11

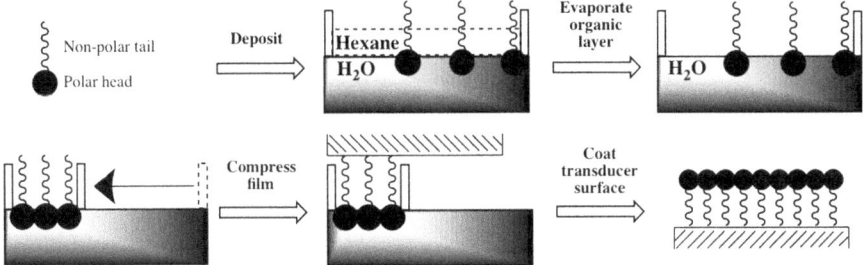

Figure 1.5 The Langmuir–Blodgett trough and deposition of an amphiphile on a transducer to produce an assembled monolayer.

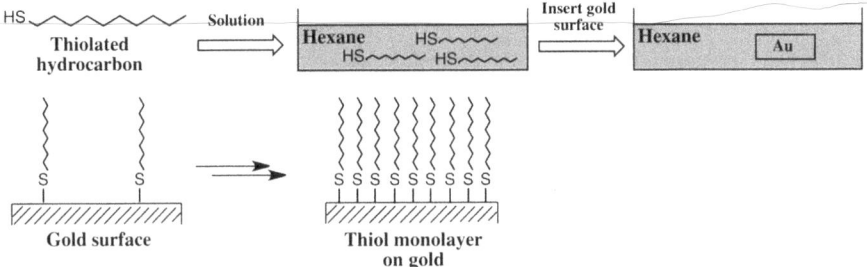

Figure 1.6 Long chain alkyl thiol deposition onto a gold substrate to generate a self-assembled monolayer.

widely prevalent as signaling devices in biology. Thus they show significant potential in the world of biosensor technology. The protein or probe employed can be incorporated into the trough preparation and transferred together with the amphiphile onto the substrate of a biosensor structure.[25,26]

An important chemistry which has been developed significantly in recent years is the self-assembled monolayer (SAM). This approach relies on the use of linking molecules which are engineered to spontaneously form ordered molecular assemblies on solid inorganic substrates. By far the most common strategy which has spawned a large literature has been the assembly of relatively long chain thiols on the surfaces of clean gold substrates.[24,27–30] The distal end of bifunctional thiols can then be employed for conventional covalent binding of proteins and oligonucleotides, *etc.* as described above[31] (Figure 1.6). This chemistry is very attractive for the attachment of biochemicals onto devices used for surface plasmon resonance and certain acoustic wave detection because these systems often employ gold in their fabrication. An alternative technique is the use of trichlorosilanes rather than thiols which can be bound to surfaces that have been functionalized with –OH groups.[32] Examples of substrates in this case are silicon dioxide (*e.g.* quartz) and indium tin oxide (ITO). This type of chemistry can be utilized in conjunction with diluents in an attempt to avoid spatial crowding as discussed above (Figure 1.7). There is considerable evidence with this method that

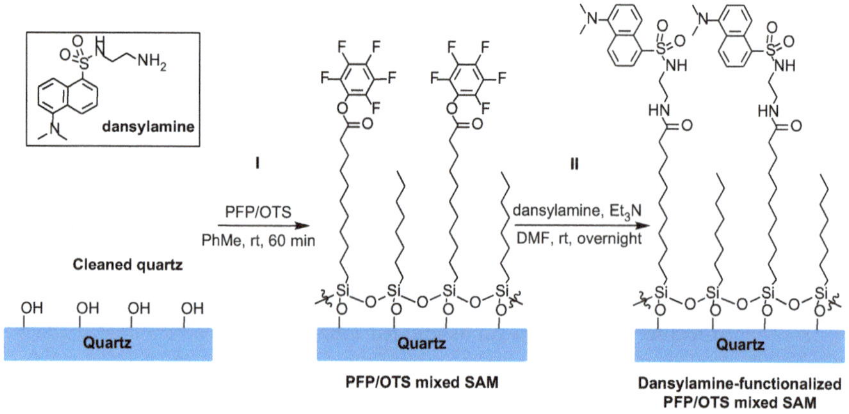

Figure 1.7 Example bifunctional trichlorosilane adlayer on quartz surface with short chain diluent in place to avoid probe crowding.
(Reprinted by kind permission of the Royal Society of Chemistry.)

non-specific adsorption can also be minimized through the use of oligoethylene-based linkers rather than simple alkyl chains.[33] This is an extension of biomaterials technology where polyethylene glycol polymer has been known for many years to enhance the biocompatibility of implants.

1.3.5 The Molecularly Imprinted Polymer

Although not strictly an immobilization technique, we mention for convenience at this point the molecular imprinted polymer (MIP).[34] This approach bears a resemblance to the entrapment methodology outlined previously. In MIP technology, created originally by Mosbach and co-workers,[35] a particular target molecule is surrounded chemically by a deliberate polymerization process in order to produce a 'mimicked' biochemical three-dimensional matrix. The final step in the whole process is the washing out of the target molecule leaving a cavity ostensibly analogous to that for antigenic species in antibodies (Figure 1.8). Presumably the cavity is capable of selectively binding the target molecule *via* a variety of chemical forces such hydrogen bonds and van der Waals forces orchestrated in a three-dimensional configuration. The technique has been used in combination with several types of device, although the instigated selectivity often appears modest. It is generally the view that an element of 'collapse' of the cavity is responsible for the reduction of selectivity. However, this is an active research field and no doubt considerable effort will be made to enhance selectivity with respect to MIP technology. One such approach is that of the notion of 'clonal selection' which has been employed highly successfully in nucleic acid and protein microarray technology. In this scenario no attempt is made to design a particular cavity. Instead a large array of variable cavity polymers is deposited onto a substrate and all are allowed to interact with a particular analyte. As for the aforementioned biochemical

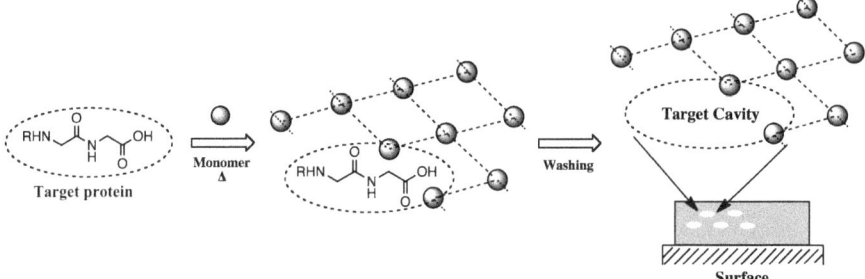

Figure 1.8 Molecularly imprinted polymer. Schematic of proves where polymerization is conducted around protein, followed by removal to produce a recognitive cavity.

technologies, the 'right' polymer can be found by interrogation ready for further development.[36]

1.4 Device Transduction of Biochemical Interactions

A number of sensor strategies have been developed over the years for the purpose of transducing a variety of biochemical events into an electrical signal. Here we review concisely the three main categories—electrochemical, acoustic wave and optical approaches. It should be noted that this compendium is not exhaustive; for example, there are pyroelectric and other devices available. However, the areas specified above dwarf all other sensors in terms of numbers of applications.

1.4.1 Electrochemical Systems

There are a number of different approaches possible from the world of electrochemistry, which involves the transfer of electrons at a solid–liquid interface. A précis of these is presented in Table 1.2. A particularly useful feature of many of these techniques is that they can be combined 'naturally' with various forms of integrated electronic circuitry. Indeed the necessary chemistry can be imposed directly on the surface of such an electronic device (see later).

The literature displays a number of reviews of this transduction which can be accessed by the reader for more detail.[37–40] Here we deal with a concise compendium of key characteristics, systems and devices.

1.4.1.1 Potentiometry

The defining principle of sensors based on potentiometry is that an indicating electrode for the analyte of interest is incorporated in an electrochemical cell (with reference electrode) for which minimal or no current is passed. In this category the ion selective (indicating) electrode (ISE) forms the basis of systems capable of detecting species of biochemical interest.[41,42] By far the most common one in use is the glass electrode, which is sensitive to changes in hydronium ion concentration. This type of electrode is often termed a membrane electrode in view of the development of a junction potential connected to selective ion exchange processes at the membrane-solution interface. Other systems available for ion selective detection include electrodes which employ crystalline matrices, *e.g.* LaF_3 for F^-, and those which incorporate liquid ion exchangers in polymer matrices such as systems for sensing Ca^{2+}. A typical general equation that gives the potential, E, of an electrode for an ion, a, in the presence of an interfering species, b, is:

$$E = E^0 - (RT/xF)\log(a_{a} + K_{a,b}a_{b}) \tag{1.2}$$

where R is the gas constant, T is temperature, F is the Faraday constant, x is the number of electrons involved in the electrochemical process, a_a and a_b are the activities of the two ions, and $K_{a,b}$ is a selectivity coefficient for the electrode.

The ISE has been the basic structure used to develop enzyme-based molecular sensors for a number of years.[42] In this type of device the analyte is allowed to interact with an immobilized enzyme resulting in conversion to a product which can be detected by the electrode. The product of the catalyzed reaction can be a gas such as ammonia or a cation such as the hydronium ion. In either case the electrode actually detects a change in pH in terms of response. A simple example of this approach is the electrode for sensing blood urea *via* the enzyme, urease (Figure 1.9). The product of the reaction, NH_3, alters the localized pH which is then detected by a glass electrode.

An alternative transducer to the conventional electrode outlined above is the field-effect transistor (FET), which was introduced by Bergveld[43] in the 1980s for ion sensing. Since this device can be integrated into various circuits in a miniaturized format and has been employed to a significant degree with

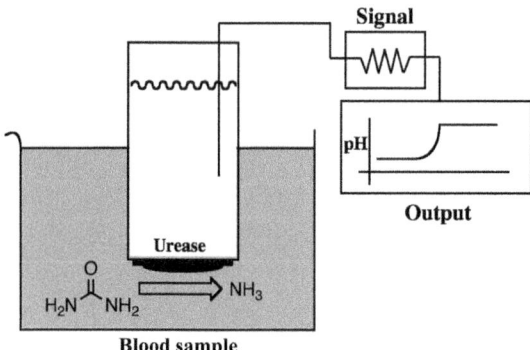

Figure 1.9 A urease-based enzyme electrode. Note device detects changes in pH as a consequence of enzyme activity.

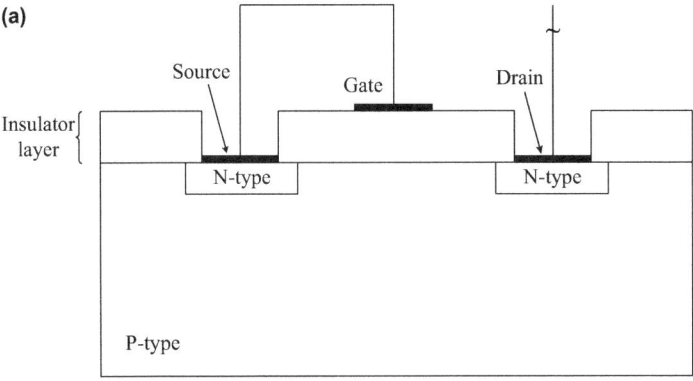

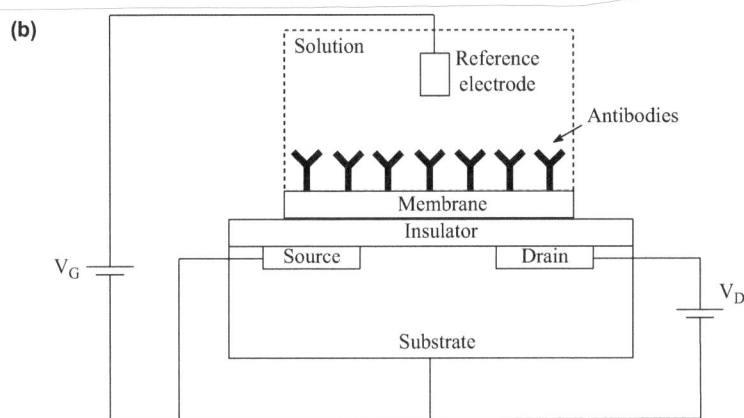

Figure 1.10 (a) The n-channel-enhancement mode insulated-gate field effect transistor (IGFET). (b) An IGFET typically set up to act as an antibody-based biosensor. Note insulating protection of source and drain contacts.

cells and neurons, we will outline its operation and functional characteristics. Figure 1.10a shows the structure of a typical n-channel-enhancement mode insulated-gate field effect transistor (IGFET). Two separate regions of n-type semiconductor materials are embedded in a p-type substrate. A conductive pathway, termed a channel, can be instigated between the n-type regions. The device is covered with an insulating layer of silicon dioxide. Using photolithography, openings are fabricated to allow electrical contact with the n-type regions. Electrical contact is also formed at the substrate surface and on top of the insulating region. One of the n-type contacts is called the source and the other the drain; a potential is applied which drives current in the channel. The contact on the surface of the center part of the insulating layer is called the gate. It is in this location that the device is employed to detect charge changes connected to chemistry conducted at the gate-solution interface. Note that certain regions of the device must be encapsulated and the system as mentioned above needs an adjunct reference electrode. The device has been employed in various formats including the detection of ion concentrations (ISFET)[44,45] or immunochemical interactions (IMMUNOFET, Figure 1.10b).[46] The imposition of such chemistry on the gate surface follows a somewhat similar approach to methods used for conventional electrodes. In addition, the strategies outlined above for biochemical immobilization are generally employed in terms of attaching proteins and nucleic acids to the gate region.

1.4.1.2 Amperometry

Potentiometry does not involve transfer of charge at electrodes since there is a mandated null-current. As mentioned above, the equilibrium potential of a particular electrode is given by the famous Nernst equation (eqn (1.2)). Alteration of an electrode potential with resulting disturbance of the equilibrium condition will yield a transfer of charge and electron movement across the liquid–electrode interface. In general terms, techniques that are based on measurement of the resulting current *versus* applied potential are called voltammetric methods. Anodic stripping and cyclic voltammetry are two examples of such techniques. Amperometry has many definitions in the literature but is more often than not simply seen as involving the measurement of current associated with an oxidation–reduction reaction at an electrode.

As mentioned above, the beginning point of biosensor technology was the oxygen-mediated electrode for the amperometric assay of glucose. Over the years the electrochemical detection of this analyte, given its importance, has been the subject of literally thousands of studies, the history of which has been beautifully reviewed by Wang.[47,48] One of the most important advances in the technology was the introduction of organometallic mediators such as ferrocene. This iron π-arene complex composed of a cyclopentadiene sandwich of Fe acts as a convenient electron acceptor for glucose oxidase, thus avoiding the dependence on the role of O_2 (Figure 1.11). The mediator has a low redox potential and can be conveniently immobilized on an electrode surface with the enzyme. Years of further research have seen a number of alternatives employed for same purpose, including derivatives of ferrocene, and efforts to improve the proximity of the system to the enzyme and electrode. All this research has contributed to the development of the point-of-care test strips for glucose assay

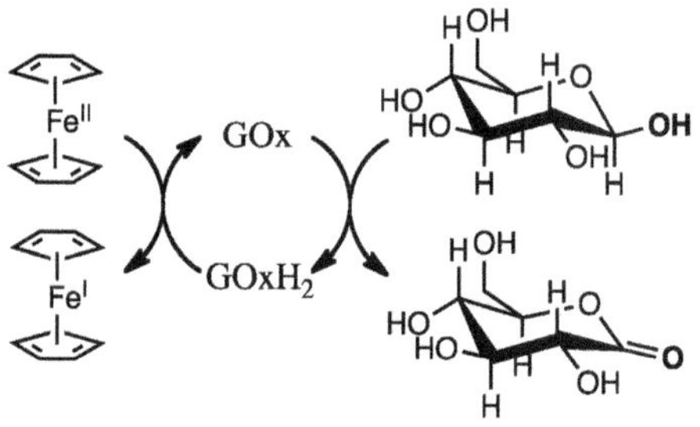

Figure 1.11 Electron mediator for glucose oxidase.

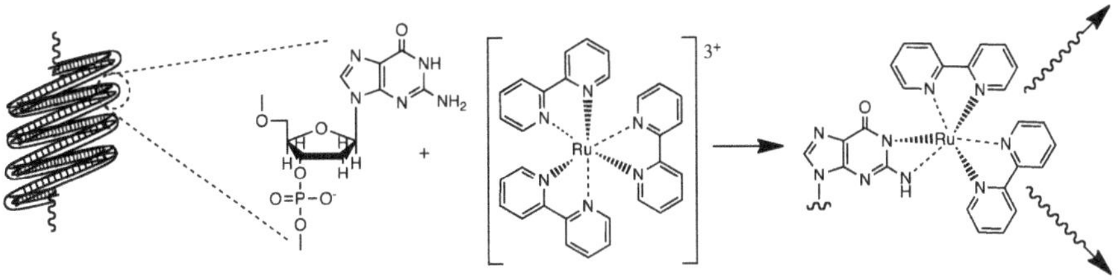

Figure 1.12 Ru(bpy)$_3^{2+}$ label of nucleic acid strand as a precursor for electrochemical detection of duplex formation.

so familiar to the diabetic patient. A further crucial aspect of all this work has been, and continues to be, the possibility of using an electrochemical device for *in vivo* assay of glucose in blood in an artificial pancreas scenario. However, despite enormous effort over the years with the corresponding financial implications, such a system is not available. The reason for this undoubtedly lies in issues concerning sensor fouling in blood and on-the-fly calibration.

The detection of nucleic acid interactions and duplex formation at an electrode surface by an amperometric protocol has been the subject of intensive research for some time.[49] One example among several is use of the redox properties of, for example, Ru complexes to enhance the electroactivity of nucleic acid species (the intrinsic redox behavior of nucleic acid moieties is insufficiently sensitive for analytical purposes). The coupling of Ru(bpy)$_3^{2+}$ mediated oxidation of the base guanine is said to lead to nucleic acid target detection at least the attamole concentration (Figure 1.12). Curiously, the electrochemistry community often describes this protocol as a 'label-free' detection strategy which is obviously invalid given the necessary use of the Ru adjunct agent. In more recent times scientists at the University of Toronto have engineered this electrochemical approach into a hand-held device, ostensibly to be employed for point-of-care assay of cancer.[50]

Finally, with respect to amperometry, we mention the technique of chronoamperometry, which involves the application of a square-wave potential to the indicating electrode with measurement of the resulting steady-state current over time. In this scenario, the concentration of an electroactive species close to the electrode surface is reduced to zero. Accordingly, for electrochemistry to take place the entity of interest must be transported to the device by diffusion. The technique was employed some time ago to examine neurotransmission and the electrochemical behavior of slices of brain tissue.[51]

1.4.1.3 Impedance spectroscopy

Electrochemical impedance spectroscopy (EIS) has been available for more than 30 years. In this technique a sinusoidally varying voltage is applied to an electrochemical system and the resulting current is measured, usually over a range of angular frequencies. Recording of these parameters can be employed to compute the real and imaginary components of the electrical impedance (Z).

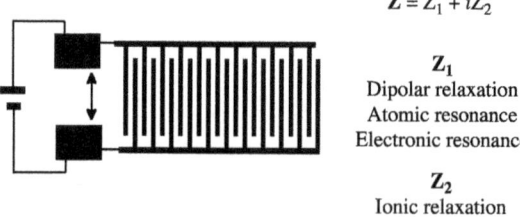

Figure 1.13 Impedance devices where chemistry is conducted in an interdigital electrode gap.

Primarily the system has been used to detect biomolecular recognition events taking place at an electrode, much as the situation with the other electrochemical arrangements outlined above.[52–54] One interesting variation on this theme was the detection of electrical changes in the gap between interdigitated electrodes associated with immunochemical interactions.[55] As can be seen in Figure 1.13 changes in the gap conductivity and capacitance are connected with alteration of the real and imaginary impedance components, respectively.

In a similar fashion, EIS has been employed to detect nucleic acid duplex formation in the gap between interdigitated electrodes. The signal in this case has its origins in the conductivity differences in the gap associated with single and double strand DNA species.[56] An attractive feature of this technology is its capability to be exploited in tandem with nucleic acid microarray assays. In this method a large array of single strand oligonucleotides or DNA are exposed for hybridization with a sample containing possible target complementary strands. Normally, the gold standard for the detection of the various levels of duplex formation is confocal fluorescence microscopy, which requires the use of luminescent labels.

1.4.1.4 Electrochemical Detection and the 'Nano' World

Science has witnessed an explosion of relatively recent interest in nanotechnology and this has found its way into electrochemical research. Devices have been constructed from nanowires,[57–59] nanorods, nanoparticles and nanotubes in addition to 'nanofilms' deposited on electrode surfaces. Much of this activity has been concentrated on nanowires where the diameter of the structure is on the nanometric scale, although the length dimension may well be much greater. The main property of interest with respect to such devices is defined by the very significant surface-to-volume ratio for nanowire atoms compared with more macroscopic systems. Accordingly the Debye screening length is expected to penetrate the width of the wire to a large extent.[60] This leads to the consideration that conduction in the wire will be controlled by quantum state effects. Accordingly, any changes in surface charge, such as instigated by biomolecular interactions, will result in highly sensitive changes in wire conduction. Among several attempts to exploit this projected high sensitivity has been work on nanowires used in tandem with FET technology.[60,61] Such a configuration, in principle, allows multi-analyte sensing as depicted in Figure 1.14.

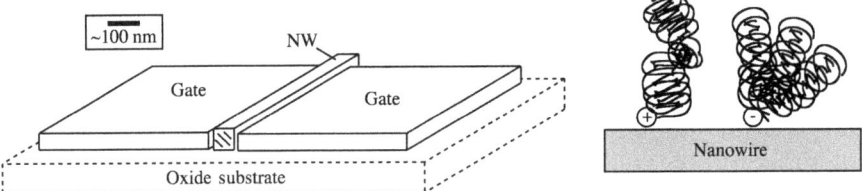

Figure 1.14 Basis of detection based on combination of a nanowire and field-effect transistor.

Although offering attractive possibilities in terms of potential device sensitivity and miniaturization, it remains to be established if issues of convenient biochemical attachment and overall incorporation into required circuitry can be solved from a routine perspective. There is also the ever present specter of how such structures would be expected to behave if employed in complex media such as blood. The enhanced sensitivity referred to above may actually represent an Achilles' heel unless the interfering surface effects can be minimized or eliminated.

1.4.2 Acoustic Wave Physics and Devices

The inception of the use of acoustic wave devices for the detection of biochemical interactions is generally regarded to have taken place in the 1980s. Development of the technology has had a checkered history in that controversy surrounded earlier research efforts. The first attempt with a surface acoustic wave (SAW)[62] device for immunochemical detection was later shown not to involve Rayleigh waves at all,[63] and around that time it was commonly believed that the thickness shear mode (TSM) device could not be used in fluids because of severe damping effects. Despite this notion, Konash and Bastiaans[64] were able to use a bulk acoustic wave device as a liquid chromatographic detector by exposing only one face of the seminar to fluid. Moreover, with respect to the latter, the invalid concept that the TSM operating in liquids is exclusively a 'mass' sensor persists to modern times.

Acoustic wave sensors are very much associated with piezoelectric physics. However, a number of structures are available which do not employ piezoelectric components in a direct sense. Accordingly, the latter are included for completeness in this comprehensive review of the technology.

1.4.2.1 Essentials of the Piezoelectric Phenomenon

An excellent definition of piezoelectricity has been provided by Cady,[65] '*piezoelectricity is the electric polarization produced by mechanical strain in crystals belonging to certain classes, the polarization being proportional to the strain and changing sign with it*'. We now recognize that the reverse is true also, that is, a specific crystal can be mechanically strained when subjected to an electric polarization. The latter phenomenon is often referred to as the converse piezoelectric effect. Historically, the Curie brothers are universally credited with the discovery of piezoelectricity as a result of their pioneering work on crystals at the Laboratory of Mineralogy at the Sorbonne in Paris. Lord Kelvin[66] subsequently worked on a thermodynamic theory of the effect and also late in the 19th century another giant of the scientific world, Lord Rayleigh,[67] showed that surface waves could be instigated in elastic solids. This work was the genesis of surface-launched waves in piezoelectric materials discussed briefly in this text. In 1895 Marie Sklodowska married her supervisor Pierre Curie and in their work on the discovery of radium produced what was termed a 'quartz electric balance' as stated in her fascinating thesis.[68] Although there have been a number of claims as to the initiation use of piezoelectric devices as chemical sensors, it is certain that the work described in this thesis is genuinely the first.

Any description of the piezoelectric effect must begin with the concept of viscoelasticity displayed by certain crystals and solids. The application of a stress on a viscoelastic medium can be described by a generalized tensor form of Hooke's famous equation:

$$T_{ij} = c_{ijkl} S_{kl} + \eta_{ijkl} S_{kl} \tag{1.3}$$

where T_{ij} and S_{kl} represent the stress and strain tensors, respectively. The coefficients c_{ijkl} and η_{ijkl} are, respectively, the fourth-rank elastic coefficient tensor and the fourth-rank viscoelastic coefficient tensor. In order to derive the wave equations connected to the piezoelectric phenomenon it is necessary to couple eqn (1.4) to those of electromagnetism. However, not all solids display the phenomenon so we digress briefly to outline the crystals that do indeed yield the effect.

The devices used in biosensing have in common the deformation caused in a piezoelectric crystal by the application of an electric field. The origin of this effect lies in the interaction between electric charge and elastic restoring forces in the crystal. All importantly the phenomenon cannot take place in a crystal possessing central symmetry. The piezoelectric point groups exhibiting this property with full Hermann–Maugin notation are presented in Table 1.3. Note that there is at least one piezoelectric crystal in each crystallographic system. In order to maximize the coupling of electrical and mechanical affects it is conventional for practitioners to use particular slices of crystals or cuts. For example, with respect to the major piezoelectric material, quartz, these are AT-, ST- and BT-cuts, *etc.* For interested readers the Institute of Electrical and Electronics Engineers (IEEE) notation for the various slices is given in ref. 69.

It is now possible to proceed to the coupling of Maxwell's famous laws of electromagnetism with eqn (1.5). From this combination, following certain assumptions, the important equations known as the classical piezoelectric constitutive relationships can be expressed

$$T_{ij} = c_{ijkl} S_{kl} + \eta_{ijkl} S_{kl} + e_{ijk} E_k \tag{1.4}$$

$$D_i = \varepsilon_{ik} E_k + e_{ikl} S_{kl} \tag{1.5}$$

Table 1.3 Piezoelectric point groups.

Crystal system	Hermann–Maugin notation
Cubic	23
	$\bar{4}3m$
Tetragonal	4mm
	$\bar{4}2m$
	422
	4
	$\bar{4}$
Orthorhombic	222
	2mm
	m
Monoclinic	m
	2
Triclinic	1
Hexagonal	$\bar{6}$
	$\bar{6}2m$
	6
	6mm
	622
Rhombohedral	3
	3m

Here E_k is the applied electric field, e_{ikl} is the third-rank piezoelectric coefficient tensor, ε_{ik} is the second-rank dielectric strain coefficient tensor and D_i is the electric displacement. (The superscript E recognizes that the elastic and viscoelastic coefficients are being held under a constant field E_i.) The latter parameter is connected to the concept of induced dipoles and polarization, which will be very familiar to the physical chemist. The last task in the development of an overall examination of the piezoelectric effect and device characteristics is to take account of the fact that it is more often than not the case that E_i is allowed to oscillate. This in turn leads to a set of wave equations (described in more detail in ref. 70) which link together the applied electric field and actual instigated particle displacement in a time-dependent form. Knowledge of the parameters incorporated in these equations allows rigorous solutions, which in turn, leads to the crucial possibility to quantify the response of various devices in biosensor applications.

Finally we introduce the nature of the movement instigated in the various acoustic wave devices and the role of the phenomenon of resonance. Movement of particles in a piezoelectric crystal caused by an oscillating electric-field induced stress, restored by elastic forces, leads to standing or travelling waves in the material. The actual displacement (termed 'polarization') is of three types—linear, elliptical and circular. The various devices to be discussed subsequently possess waves with shear or compressional character in operation that are constructed from these movements. As with other forms of wave propagation, the confinement of a wave within boundaries leads naturally to the instigation of the phenomenon of resonance. This involves, to a first approximation, the complete constructive interference of the standing waves mentioned above. In this connection, with respect to acoustic wave devices, the term 'piezoelectric acoustic resonator' is often employed.

1.4.2.2 Piezoelectric Bulk Acoustic Wave Biosensors

In this type of device, particle motion and the acoustic wave involve the complete thickness of the structure. We introduce two sensors that display these characteristics—the TSM and its sister device, the electromagnetic acoustic wave sensor or EMPAS. The TSM, because of its low cost and convenience in operation, is represented by far the most research activity and applications of any acoustic wave system; accordingly, it is dealt with in more detail than other configurations.

The TSM is composed of an electroded (often gold) piezoelectric wafer (usually AT-cut quartz) (Figure 1.15). An oscillating electrical potential is applied *via* the electrodes in order to drive mechanical motion in the device. A resonant acoustic shear wave is generated which travels through the piezoelectric material with little energy dissipation. The wave is reflected at the device–surroundings interface in order to maintain a standing wave (Figure 1.15). In chemical and biosensor applications material is generally added to the device surface, for example, species involved in a bimolecular interaction. On a simple theoretical basis this scenario was the subject of very early work by Sauerbrey[71] who showed that, when material is deposited on the device surface, the resonant frequency is changed according to his famous equation, which is expressed in eqn (1.6) for quartz as the piezoelectric material:

$$\Delta f = -\frac{2f_0^2}{A\sqrt{\rho_q \mu_q}} \Delta m \qquad (1.6)$$

where Δf is the frequency change, f_0 is the primary resonant frequency of the sensor, A is the effective surface area associated with the piezoelectric process, ρ_q

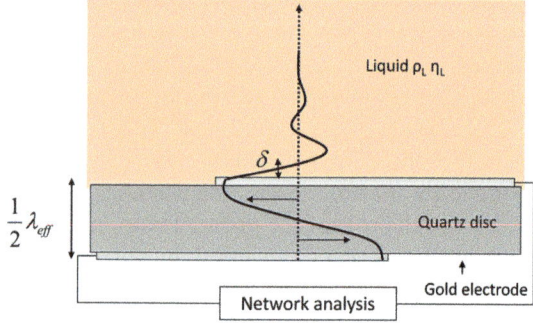

Figure 1.15 Thickness–shear mode acoustic wave device and standing wave in device with propagation of a damped acoustic shear wave into liquid.

is the density of quartz, μ_q is the shear modulus of the particular cut of quartz employed and Δm is the change in mass. Key assumptions in this treatment are that acoustic energy is confined to the device and film in place on its surface, and that the acoustic properties of the film are identical to those of quartz. Given the fact that the film in place under these conditions predicts an extension of wavelength and, always, a reduction in resonant frequency, the device has been employed universally as a sensor of added (or lost) mass. This notion has spawned the ubiquitous term 'quartz crystal microbalance' or QCM. Unfortunately, this argument has been widely extended to operation of the device in the liquid phase and has become something of a dogma, especially in the community of electrochemists. In such a medium, it is a reality that acoustic energy is transferred to the surrounding liquid resulting in a damped wave *via* the effects of bulk phase viscosity (Figure 1.15).

The years have seen numerous attempts at the development of theories for operation of the TSM in liquids (reviewed in ref. 72). These have included the influence of liquid density, conductivity, dielectric constant and viscosity, and interfacial properties such as roughness, slip and stress. A prominent earlier model was that provided by Kanazawa and Gordon[73] which predicts that the change resonant frequency will be proportional to the square root of the liquid viscosity/density product.

Since the introduction of the earlier models, it has now become apparent that acoustic coupling effects and the phenomenon of interfacial slip at the liquid–solid interface are important properties which, in part, govern the response of the device when operated in the liquid phase.[74–76] One particularly crucial aspect of these phenomena, in sharp contrast to the QCM notion, is that imposed mass on the device surface can lead to frequency increases under certain conditions. Just as important, the implied perturbation of acoustic coupling by interfacial chemistry offers a unique mechanism for biomolecular detection. This includes the detection of phenomena such as biomolecule conformational changes.

Experimentally, a number of approaches have been employed to measure the resonant frequency and other acoustic parameters, in some cases in a static fashion, and in others, in flowing liquid through a flow-injection configuration. The most rigorous approach is that provided by what is often termed acoustic network analysis. In this method the magnitude and phase of impedance of the sensor are determined at a set of frequencies under resonance conditions. A network analyzer is employed to record the Butterworth–Van Dyke equivalent circuit which essentially relates physical properties of the device to electrical parameters (Figure 1.16). Note that the true measured series resonance frequency is detected at zero phase angle. Other determined parameters from the equivalent circuit are the motional resistance, R_m, which is importantly associated with energy dissipation in the system including any added surface film, inductance, I_m and two capacitances, C_m and C_0. The second approach, which is available on a commercial basis, involves the measurement of a resonance frequency and an energy dissipation factor, D.[77] The latter parameter provides analogous information to the R_m factor

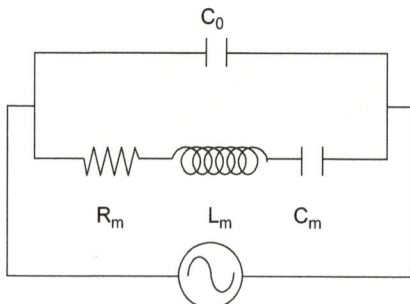

Figure 1.16 Electrical equivalent circuit for operation of a thickness shear mode acoustic wave sensor. Depicted are motional resistance (Rm), inductance, (Lm), and circuit capacitances (Co and Cm).

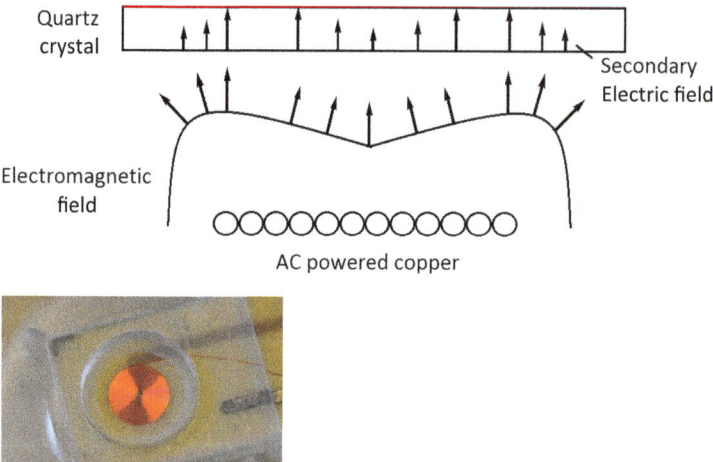

Figure 1.17 Electromagnetic acoustic wave sensor. Acoustic resonance is induced in electrode free quartz by secondary electric field from flat-spiral coil.

mentioned above. There is also the possibility with this system of measuring the frequency overtones of the device.

In terms of applications, recent years have seen a rapid increase in the development of TSM technology which has been employed to detect surface-induced protein conformational changes,[78] immunochemical interactions,[79] nucleic acid hybridization[80] and cell–surface attachment.[81] As mentioned above, the major reason for these advances is that the response of the TSM in liquids is extraordinarily sensitive to interfacial phenomena, which include surface free energy, charge and viscosity effects.

A recent acoustic wave device development is the introduction of the EMPAS structure.[82] In this technology, acoustic waves are instigated in an electrode-less quartz wafer by an electromagnetic field generated in close proximity to the wafer by a flat spiral coil (Figure 1.17). The secondary electric field associated with the coil drives the piezoelectric effect in the device. The configuration possesses a number of important advantages.

- It is not necessary to operate the device with a metal electrode in place and with electrical connections. This renders advantages in terms of flow-through design.
- Crucially, it is possible to operate the sensor at ultra-high frequencies (*e.g.* 1 GHz) *via* bulk acoustic wave overtones. This leads to higher analytical sensitivity.
- It is possible to tune the device with ease to specific frequencies, which could potentially lead to important interfacial chemical information.
- The surface chemistry for biomolecule attachment involves SiO_2 which is a much higher developed area of chemistry than is the case for binding to metals such as gold.

The system has been applied to the detection of human immunodeficiency virus (HIV) based antibodies and bacteria.[83]

1.4.2.3 Piezoelectric Surface-launched Acoustic Wave Biosensors

A number of devices in this category have been described in the literature together with a rather limited number of applications to bioanalytical chemistry, at least as compared with the array of those connected to the TSM. 'Surface-launched' in this context means that acoustic waves are generated in a piezoelectric substrate by transducers that are placed on the surface of the material. Particle movement, which is generally detected by separate tranducer(s) imposed on the same surface, is often restricted to the 'near' surface of the substrate, unlike the case for the TSM discussed previously. For an introduction to several of these devices see ref. 84.

The first device to be considered is the surface acoustic or Rayleigh wave (SAW) sensor which is, conventionally, an extremely common component of microelectronic and integrated circuitry. We introduce this device since it is very much the forerunner of the SAWs in use in the sensor field. In this structure an interdigital transducer (IDT) is fabricated on the piezoelectric substrate (*e.g.* ST-cut quartz) as shown in Figure 1.18. The effect of this arrangement is to create acoustic waves near the surface of the substrate which possess a wavelength far smaller than the overall thickness of the structure. As seen in Figure 1.18, the particle motion of a Rayleigh wave contains both shear and compressional components and travels in an elliptical path. In the delay-line configuration, the IDT to the right side of the device detects the arrival of the acoustic particle motion. (There are other designs of the SAW structure.) The chemistry related to sensing processes is conducted on the area between the IDTs and it is generally the case that the electrodes have to be isolated from liquid if the device is operated in that medium, in a similar fashion to that described above for FETs. However, earlier work has shown unequivocally

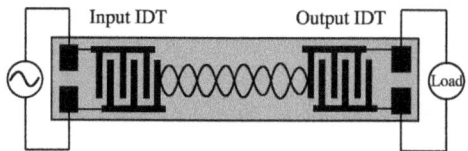

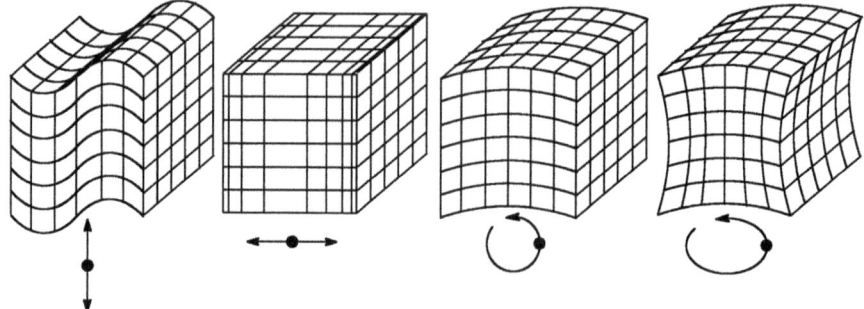

Figure 1.18 Schematic of surface acoustic wave device. Rayleigh waves contain both shear and compressional components.

that, in the liquid phase, the compressional component is heavily compromised by attenuation effects rendering its operation untenable.[63]

Historically, the result with the Rayleigh wave device has led to the use of devices which possess shear wave components only, but that are induced by surface launching. Examples of various devices employed *via* this technology are shown in Figure 1.19. One of the earlier structures is the shear-horizontal surface acoustic wave (SH-SAW) device, for which the particle displacement is depicted in Figure 1.19.[85] This device was followed by surface transverse wave (STW; see Figure 1.19)[86] and Love-wave[87] sensors. The use of these biosensors, which all involve very similar biomolecule immobilization strategies to those outlined above, have been reviewed by Rapp and coworkers.[88] Finally, we note that there are also a number of structures where, although electrodes are fabricated on the surface of a device, the complete structure is caused to vibrate. The sensors are generally termed 'plate mode' devices (see ref. 70).

Although many of these devices have been employed very successfully in gas phase sensing, and exhibit relatively high frequency of operation conveying high sensitivity, applications of the TSM as a biosensor still dwarf those of SAWs in numbers.

1.4.2.4 Non-piezoelectric Acoustic Wave Devices

In this section we comment briefly on devices which display potential for chemical and/or biosensing but which do not contain a piezoelectric substrate

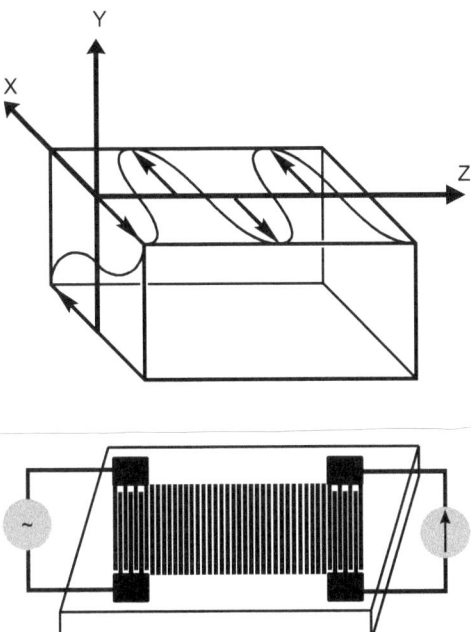

Figure 1.19 Particle movement for a shear-horizontal acoustic wave device and schematic of a surface transverse sensor.

as the main transducer of imposed chemistry. The first of these is the magnetic acoustic resonator sensor or MARS device.[89] In this technology, the magnetic direct approach is employed to instigate acoustic waves in a non-piezoelectric substrate such as glass. In an analogous fashion to the EMPAS system described above, a flat spiral coil, with a radiofrequency (RF) current is used in conjunction with a permanent magnet to induce radial acoustic waves in an aluminum film present on the surface of the glass substrate. In summary, the metal film 'drags' the substrate into vibrational motion and a change in behavior of imposed chemistry on the glass surface can be detected to generate a response. Although available for a number of years this acoustic sensing mechanism has not seen widespread use in bioanalytical chemistry, probably because of difficulties in obtaining acoustic resonance on a routine basis and the relative ease of operation of the EMPAS device which does not require the use of an adjunct magnet.

Finally, we mention the instigation of acoustic waves in thin rods of non-piezoelectric material.[90] In this arrangement an ultrasonic longitudinal wave is piezoelectrically excited in a glass horn which in turn is coupled to a thin rod. Extensional, flexural or torsional acoustic waves can be induced in the rod and are detected by a second transducer in a delay-line configuration similar to that mentioned above for the SAW. Although not used directly for biosensing, the system possesses significant promise in view of the modern advances in nanowires and nanorods alluded to above.

1.4.3 Electromagnetic Radiation: Optical Devices

Virtually the full gamut of physics offered by optical science has been employed over the years for the detection of fundamental biophysical processes, biochemical binding events and species of bioanalytical interest such as biomarkers for disease. Techniques include those based on measurements of absorption, luminescence, interference, reflectance, scattering (including Raman spectroscopy) and refractive index phenomena. The wide variety of techniques employed in optical sensing are reviewed well in ref. 91. The use of electromagnetic radiation in medicine (*e.g.* diseased and damaged tissue detection by infrared and Raman spectroscopies) is ubiquitous[92,93] and very exciting as the technique pertains to the characterization of tumors or tissue damage.[94–96] However, here we concentrate on optical transducers which appear most often in terms of bioanalytical sensor technology. The field from this perspective has been reviewed, especially in terms of the advantages presented by the possibility for label-free detection.[97] In many cases there is a specific focus in view such as the interesting discussion of clinical diagnostic technology.[98]

In contrast with conventional instruments for optical spectroscopy, the delivery of radiation in biosensor technology takes an alternative pathway and often involves alternative structures. Accordingly we begin with the 'confinement' of light energy.

1.4.3.1 The Optical Wave Guide and Fibers

At the heart of a number of biosensor optical devices is the optical fiber, a structure ubiquitous in the world of communications technology. In essence, light transmission along the fiber is produced *via* a guided wave through an integral process of internal refection. As depicted in Figure 1.20, light can transmit along a fiber with complete internal reflection or *via* some 'loss' of electromagnetic energy though the effect of reflection/refraction at the interface where materials of different refractive indices are involved. The physics behind these processes are elegantly described in the beautiful classic text by Lisa

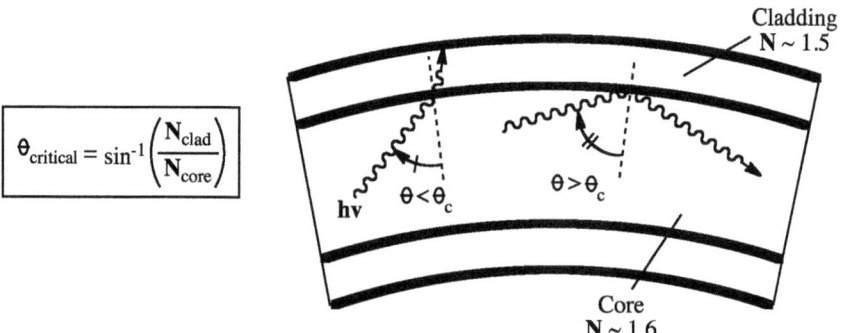

Figure 1.20 Segment of optical fiber depicting conditions for reflection of radiation at the internal interface.

Hall.[99] Both these processes have been employed extensively in the development of biosensors and in this context the term 'optrode' has appeared, as spawned by the older concept of an electrode in the field of electrochemistry. In the first methodology often termed the 'extrinsic' device, the fiber is simply used as a delivery system for light interaction with an optical configuration place at the distal end of the structure. Unsurprisingly, this arrangement has been employed, for example as the remote correlation of conventional, Beer–Lambert absorption photometry,[100,101] (Figure 1.21(a)). In the earlier days, the advantages of this sort of configuration were stated as lying in the possibilities for removal of the sensing area from the source, correction for variation of the latter and the introduction of a convenient reference system. These features have, for some time, been considered to be superior to some aspects of electrochemical devices such as those described above.

In the second scenario called the 'intrinsic' structure, some light energy finds its way under certain conditions into the medium outside the fiber in the form of a penetrative evanescent wave (Figure 1.21(b)). Importantly, the interaction of surface chemistry at this interface with the evanescent wave can result in perturbation of the light phase, intensity and polarization. There are many examples of such an arrangement in biosensor detection.[102–104] One such elegant case is the fluorescence-based system pioneered by Krull and his various co-workers[105,106] at the University of Toronto at Mississauga. Here, for example, nucleic acid duplex formation at the fiber–liquid interface can be monitored very sensitively though various aspects of luminescent spectroscopy (Figure 1.22). This type of device had been employed successfully in the assay of 'real' samples such as detection of pathogens.[107]

Finally, various configurations of optical fiber based systems have emerged generated with the overall aim of increasing sensitivity (*e.g.* Figure 1.23). These are summarized nicely in ref. 108.

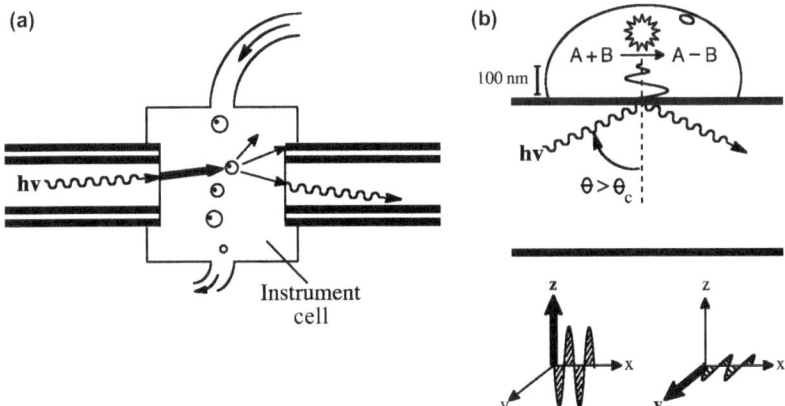

Figure 1.21 (a) Schematic of extrinsic fiber optic delivery of radiation to a cell for absorption measurement. (b) Evanescent radiation penetrating to the exterior of a fiber in an intrinsic configuration.

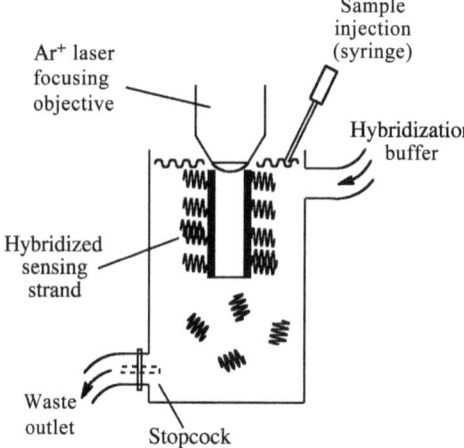

Figure 1.22 Detection of nucleic acid hybridization on surface of optical fiber *via* excitation with evanescent radiation.

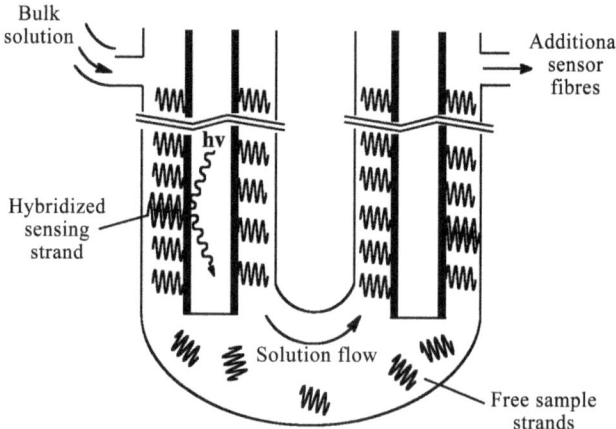

Figure 1.23 Multiple optical fibers can be employed in order to enhance sensitivity of detection.

1.4.3.2 Surface Plasmon Resonance

This technique has very often been regarded as representing the most profound commercial success in terms of biosensor technology, with the possible exception of the glucose electrode. From its genesis at the University of Linköping in Sweden in the 1980s, the initial commercial sale of various surface plasmon resonance instruments has changed hands several times and also has been the subject of strong competition. The technique as applied to biochemical detection science has been reviewed a number of times.[109–115]

To the non-specialist the physics of the generation of surface plasmon waves is somewhat difficult to comprehend. At the heart of the effect is the

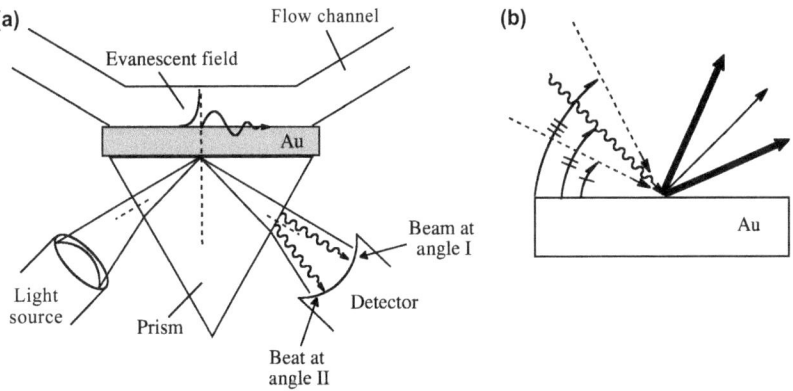

Figure 1.24 (a) Kretschmann configuration for observance of surface plasmon resonance. (b) Incident angle *versus* reflected intensity of radiation as a result of surface plasmon resonance.

observation that valence electrons of metals such as gold and silver exhibit oscillations in their density. If these oscillations, in the form of surface waves, are present at the interface of the metal with a material of a different dielectric constant, they can be excited by the introduction of an agent such as electromagnetic radiation or an electron beam. The classical approach to the study of this phenomenon was initially introduced by Kretschmann[116] and is shown schematically in Figure 1.24(a). The incident radiation is reflected at the metal–dielectric boundary resulting in an evanescent wave in the metal, much as described above for optical fibers. At the correct (resonant) angle this wave can couple in a resonant fashion with the frequency of the incoming radiation, resulting in the excitation of electron density oscillation mentioned above, leading to the term surface plasmon resonance (SPR). (In this context, the reader will be familiar with nuclear magnetic resonance and, indeed, the acoustic wave devices described above function *via* an analogous coupling of electrical with mechanical energy.) In its simplest form this process can be described on a physical basis by the following equation:

$$2\pi/\lambda \times n_p \sin \theta = 6_{sp} \qquad (1.7)$$

where λ is the wavelength, n_p is the material refractive index, θ is the angle of incidence of the radiation and 6_{sp} is a parameter called the propagation constant.

On an experimental basis, the SPR process can be affected by modification of the metal surface (such as attachment/loss of biomolecules), which is associated with a change in the dielectric constant of the surrounding medium. The surface alteration is generally explored through detection of shifts in the SPR angle, observed wavelength of absorption, or change in the position of reflectivity. With respect to the first of these cases, which represents a common approach, alteration of the angle of incidence of the incoming light with concomitant monitoring of the intensity of the reflected light intensity is monitored. A resonant transfer of electromagnetic energy into the surface plasmon wave is observed through a minimum in the plot of incident angle *versus* reflected intensity (Figure 1.24(b)). The key feature experimentally is how well a specific configuration is capable of sensing a change in the intensity of the reflected light, often termed RIU. Most commercial instruments have a limit of detection of around 10^{-6} to 10^{-7} RIU.

A simplified schematic of a typical SPR experiment is shown in Figure 1.25. This type of instrument designed for work in the biosensor arena is generally capable of analyzing several channels (gold surfaces) using microfluidic sample introduction in a real-time, label-free fashion through flow-injection technology. In certain cases, for the purpose of biomolecule attachment to the sensor 'chip' and for enhancement of sensitivity, a dextran layer some 100 nm thick is imposed at the device–liquid interface. The system generates a response plot, which is often referred to in the field as a 'sensorgram', presumably in light of chromatograms and the like! An example of SPR-based plots is depicted in Figure 1.26. With respect to applications, the SPR technique has been employed in a wide variety of cases such as the adsorption of proteins, cells and nucleic acids on gold and modified metal surfaces, and the detection of biomolecular interactions such as found in immunochemistry and nucleic acid duplex formation.[117–122] The method has proven to be particularly helpful for epitope determination in the former area.

Other instrumental arrangements for observing SPR have appeared in the literature. One example among several is the waveguide coupling of photons with plasmons. In one example of this arrangement, plasmons on a surface of a thin metal film are excited by a Gaussian-like leaky mode of an effectively single mode photonic crystal waveguide.[123] This allows efficient phase matching with plasmons at any wavelength of choice while retaining highly sensitive response to changes. Another important advance in terms of an additional form of plasmon resonance detection is SPR imaging, sometimes referred to as SPRi, which was introduced in the 1980s.[124,125] At that time it was shown that

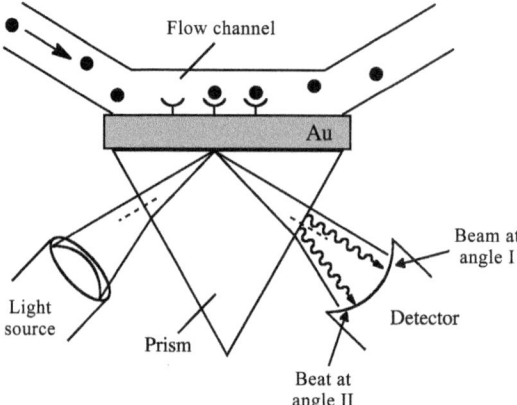

Figure 1.25 Typical SPR experiment involving flow injection on target into flow cell.

Introduction to Biosensor Technology

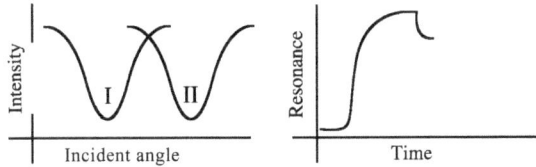

Figure 1.26 Detection *via* alteration of incident angle and production of SPR sensorgram.

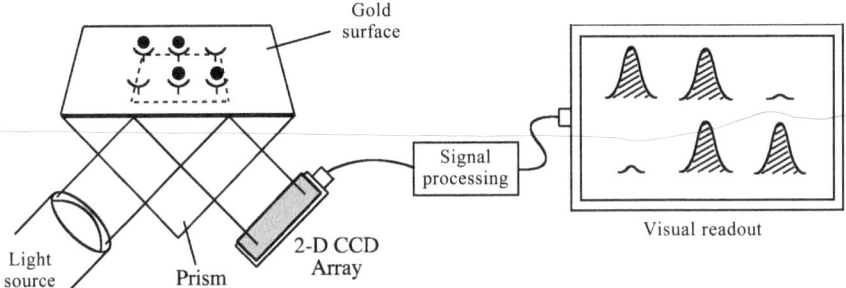

Figure 1.27 Simplified schematic of SPR imaging experiment.

laterally structured coatings on thin metal films could be examined using imaging technology. The sample is illuminated with an expended beam of parallel light over relatively large areas. With the angle of incidence and wavelength kept constant, the detection mode is the measurement of reflected light by a charge-coupled device (CCD) camera. A schematic of this sort of experiment is depicted in Figure 1.27. Since the 1980s there has been a proliferation of array-based SPR detection systems largely spawned by the concomitant rapidly developing interest in protein and nucleic acid microarray analysis in the early 2000s. These days equipment is available for the assay of thousands of domains imposed on a substrate, much as is evident with the more conventional confocal fluorescence microscopy technology. Further developments in SPR detection have been connected with improvement of chip design, use of interferometry, replacement of gold as the plasmon substrate and use of nanowire technology to observe plasmon processes.

Finally with regard to SPR, we briefly outline for the reader a comparison of the technique with acoustic wave (AW) physics, described above, since these methods are used for very similar purposes.[126]

- SPR is clearly far more advanced than AW in terms of instrument development especially when it comes to sample introduction and imaging, *etc*.
- Flow-through detection: Both techniques are conventionally employed in the real-time flow injection mode.
- Label-free: Both techniques offer the highly significant advantage in bioanalytical chemistry of being capable of operation in a label-free

fashion. This facility obviates the necessity for the addition of tags, as is often the case with fluorescence and electrochemical measurements.
- Attachment chemistry at the device–liquid interface: The necessity for use of gold as a substrate in SPR has meant that either dextran films or self-assembly chemistry has dominated the methodology required to bind biomolecules to the chip. AW is similar with respect to the gold electrodes in TSM technology, but offers more flexibility with Si chemistry for quartz. The dextran layer is regarded by some to be problematical in that analyte diffusion is influenced by the polymer.
- Sensing mechanism: For the AW case, transfer of energy to the liquid implies an acoustic couple between the device surface (and imposed chemistry) and the surrounding liquid. Accordingly, the measured properties of AW sensors, such as resonant frequency, are governed by a number of interfacial properties, such as surface-free energy, interfacial viscosity and charge. Therefore, these devices are exquisitely sensitive to the changes in interfacial chemistry associated with biochemical reactions instigated at the sensor–liquid interface. These observations confer possibilities to detect conformational shifts and the binding of very small molecules. For SPR, the resonance condition is extremely sensitive to variations in the refractive index immediately adjacent to the metal film, which is both a blessing and a hindrance! There is a level of debate as to the ability of SPR to detect small molecules, especially when they are attaching to an already bound macromolecule.
- Sensitivity and limit of detection: This is far from straightforward because incorrect terminology and reference points are prevalent in the literature and wide ranges of values are quoted. A monolayer of protein adsorbed to an EMPAS surface yields a change in frequency of 5000 Hz with noise in frequency being around 100 Hz. In terms of solution concentration the limit of detection (LOD) would be in the region of 10^{-11} M. Given the capability of SPR with respect to minimum value of RIU, the LOD number is generally regarded to be 10^{-12} M.

1.4.3.3 Interferometry

Interferometry is based on the superposition of, usually, electromagnetic waves, which yields information concerning the original nature of the waves. The reader will be familiar with the physics of the Michelson interferometer, which is at the heart of the Fourier transform infrared spectrometer. (With light of this wavelength, superposition can be examined through the use of a moving mirror for one of the light pathways.) In biosensing technology several types of device designs have been employed to attempt the detection of biochemical species, examples being Mach–Zehnder, Young and Hartman sensors[91] (depicted in Figure 1.28(a), (b) and (c), respectively).

The Mach–Zehnder apparatus involves two different light pathways, as is typical in interferometry, with one being subjected to passage through the ample. The laser radiation is then combined with a reference beam resulting in

Introduction to Biosensor Technology

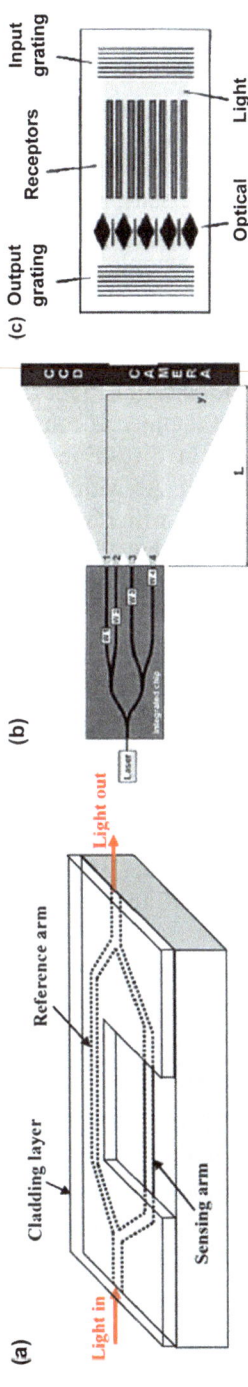

Figure 1.28 (a) Mach–Zehnder interferometer sensor. (b) Multi-channel Young's interferometer sensor. (c) Hartman interferometer sensor. (Reprinted by kind permission of Elsevier BV).

the interference-based signal. The analytical arm incorporates an evanescent field which interacts with, in this application, a biochemical moiety. Much as with SPR as described above, this process results in a change in refractive index and phase alteration which is sensed as an intensity change at the photodetector on light recombination. This technology was employed in earlier years for the detection of protein chemistry, where analytical detection limits of around 50 pM were found.[127] Since this work, several attempts have been made to improve the detection capability of the sensor such as enhancement of wave guide physics and increase of sensing surface area.[128] Although promising it does not appear that this device is being employed widely in general biosensing applications, possibly due the specialist nature of the fabrication that is required. An additional interference based structure is Young's device.[129] The overall principle is similar to the sensor mechanism employed in the Mach–Zehnder device, except that the detection mode is different. In this case the output from the sample and reference channels is combined to form interference fringes on a CCD screen. The method involves Fourier transform of the spatial intensity measure at the detector screen. As with the Mach–Zehnder sensor a number of improvements have been made with respect to design and applications of protein chemistry at the sensor surface have been demonstrated.

Finally, two other devices are the Hartman interferometer[130] (and backscattering interferometer.[131] In the former sensor, electromagnetic radiation is coupled into a waveguide using gratings, which allows the interference phenomenon to occur between a reference strip and a sample strip. As with the other two devices outlined above protein chemistry was detected as imposed on the waveguide strip. One interesting aspect of this work was that the reference strip was employed to 'remove' signals connected to non-specific binding effects. In backscattering interferometry, laser radiation is focused onto the biomolecule sample which is attached to a quite small sensing area. The reflected light intensity is detected *via* interference patterns at the detector surface. As with all the cases mentioned above the system has been used to detect the surface chemistry of proteins.

1.4.3.4 The Quantum Dot

Quantum dots (QDs) are semiconductor nanocrystals that consist of some 1000–10 000 atoms. Since their discovery in the 1980s, they have become an important component of the field what we now call 'nanotechnology'[132] rather than the more traditional, colloid science. QDs are composed of semiconductor materials with properties, because of their dimensions, that lie between a bulk substrate and individual molecules. The particles have been employed in a variety of technologies such as medical imaging, light emitting diodes, computing and photovoltaic structures for solar energy collection.[133] The potential of QDs in medical science with respect to both imaging and therapy is particularly exciting.[134–136]

QDs are fabricated from semiconductor materials such as Si, CdSe, ZnSe, ZnS and PbS. Often the dot has a core of one material with a shell of a different

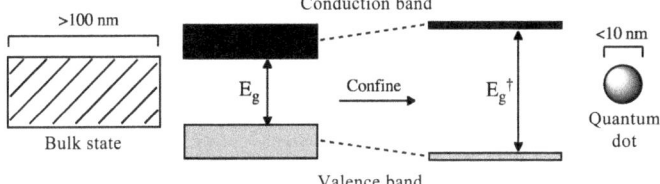

Figure 1.29 Increase in electronic band gap for quantum dot compared to bulk material.

semiconductor. They function through a principle known as quantum confinement where energy levels are dependent on the size of the particle. This physics implies a band gap between the valance electron shell and the conduction band, which is actually tunable according to the materials used for fabrication and the dimension of the dot. For a semiconductor particle of the size involved here the conventional band gaps associated with bulk materials reduce to two discrete states which are termed quantum-confined levels. The size effect leads to band gaps which are increased and, therefore, an increase of input energy is required to excite electrons to the higher level (Figure 1.29). This property in turn confers advantages on QDs with respect to fluorescence spectroscopy, which results in an important application in biosensor-like detection capability.

In contrast to the more conventional molecular fluorophores familiar to the analytical chemist, QDs exhibit superior quantum yield, longer fluorescence lifetime and the possibility for multiplexing (that is, excitation of different QDs with the same wavelength of incident light). These properties are extremely useful with respect to the detection of chemical and biochemical processes that are instigated on the high surface area of quantum dots.[137] One example is the use of the technology in a fluorescence resonant energy transfer (FRET) type of detection experiment. FRET involves the use of donor–acceptor pairs in proximity to produce fluorescence excitation and emission. In this case the QD can be employed to function in tandem with a molecular fluorophore in order to assay biochemical pairs.[137] An excellent case in point here is the detection of nucleic acid duplex formation *via* excitation of a dye with the QD-based emission; The core of the experiment is shown in Figure 1.30. Although the strategy represents an extremely sensitive detection configuration the method still requires a label, much as discussed above with respect to electrochemical detection, is subject to the usual non-specific adsorption problems if employed in complex media, and is difficult to use for time-dependent monitoring.

1.4.3.5 Raman Spectroscopy and the Metal Nanoparticle

In Raman spectroscopy a sample is conventionally irradiated with a laser light source, usually at a wavelength removed for the particular absorbance band, with the intensity of scattered radiation being measured with a spectrometer.

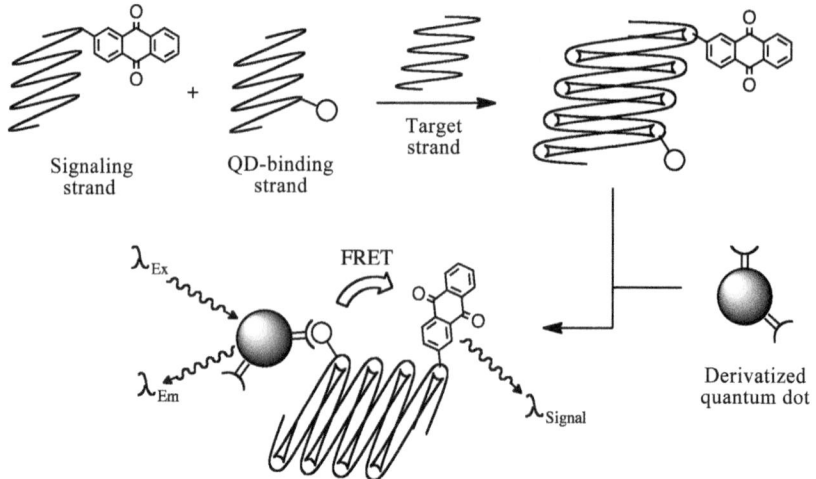

Figure 1.30 Detection of nucleic acid duplex formation *via* excitation of a dye with the quantum dot -based emission.

Scattered light produced with a lower frequency than the excitation radiation is termed Stokes scattering, whereas that with a higher frequency that that for excitation is called anti-Stokes scattering. The data obtained from the method is somewhat analogous to that found for infrared spectroscopy, *i.e.* molecular vibrational information, although the selection rules for the Raman process are different and involve induced polarizability in the target molecule. Although not considered biosensor technology, conventional Raman spectroscopy, in a wide variety of forms such as micro-Raman spectroscopy, has found widespread use in the examination of biological fluids, bacteria, cells and tissue. For several excellent examples the reader is directed to the references within ref. 93. We take one example from our own work,[96] where the technique was employed in the study of acute mouse brain injury with specific reference to the motor cortex. Tissue subject to deliberate damage showed a quite different Raman signature from a healthy brain sample. Interestingly the Raman spectra correlated with makers for apoptosis, cell death.

A system closer to the concept of a biosensor is surface enhanced Raman spectroscopy (SERS) associated with ultra-small metal particles.[138,139] SERS involves the highly significant enhancement of Raman signals, 10^3 to 10^6 in magnitude, when metal particles such as those fabricated from silver, gold or copper, for example, are employed in the Raman experiment. Much as associated with SPR described above, the SERS effect is connected to surface plasmon physics at the interfaces of, for example, two materials. The metals are formed in nanoparticles (some years ago these were considered to be in the realm of colloid chemistry!) in many different shapes. A different method to generate nanoparticles placed on substrates is the approach known as photo-or electron beam lithography (which will be very familiar to the electronic engineer).

A number of applications of the technology described above have appeared with biodetection in mind. One example is the detection of *Bacillus subtilis* spores *via* a silver-based spherical array.[140] (These spores were employed as a surrogate for anthrax spores.) Using laser excitation and spheres of dimension 600 nm on the substrate, calcium dipicolinate extracted from the spores was detected by observation of Raman spectral bands. The work was extended to anthrax spores and even the design of a portable Raman spectrometer. A second example is the detection of glucose. This experiment required the use of a partition layer (self-assembled monolayer) on the silver substrate. Glucose bands were detected following exposure to a saline solution of the molecule. In an effort to demonstrate the use of the technique for 'real' samples, the silver array was subjected to a solution of bovine serum albumin (BSA) prior to exposure to a glucose solution. Glucose could then be assayed through the use of a spectral difference technique. No attempt was made at the direct detection of the molecule in serum and, indeed, it was not clear what the advantage of the method is compared with the myriad of other methods for the determination of glucose.

1.4.3.6 Photoacoustic Spectroscopy

Finally, in this section on optical approaches to biodetection we make concise mention of the technique of photoacoustic spectroscopy, which involves the excitation of a sample with electromagnetic radiation with detection being achieved through acoustic physics. This technique has a long and distinguished history in that one of the earliest studies was conducted by Alexander Graham Bell in the last few years of the 19th Century. In essence laser light is absorbed by the sample under study which produces local heating and, in turn, a pressure wave with attendant acoustic wave generation. One of most important applications of the technology over many years has been the detection of gases such as CO_2 and NO at trace levels. Indeed a device has been described for the detection of the former molecule generated by an enzyme, which bears resemblance to an enzyme electrode for the same purpose.[141] In this case urease was employed to generate the gas, which was allowed to permeate a membrane which separated the photoacoustic cell from a solution of urea, the target analyte. It is not clear what advantage these methods would have for such determinations of enzyme activity. A further example, although not strictly biosensor technology, is work related to detection molecules associated with terrorism.[142] A quantum cascade laser has been employed with a microelectromechanical system (MEMS) designed cell to detect parts per billion (ppb) concentrations of dimethyl methylphosphonate, a nerve gas. It remains to be seen if this technology will become competitive as a method in the biosensor armamentarium.

1.4.4 Brief Summary of the Adjunct Technology Approach

To conclude this concise introduction to biosensor technology, we summarize below a number of methodologies where more than one technique is employed in a combined fashion. By far the majority of these involve a form of electrochemistry as one component.

Magnetism has been employed in conjunction with electrochemical methodology through the long-established magnetoresistive effect. This technique is concerned with the experimental measurement of the resistance displayed by a material under the influence of a magnetic field. As with the Ru complex technique outlined above, a biosensor approach in this case involves the use of a magnetic tagging agent. An example is the detection of nucleic acid hybridization on the surface of a magnetoresistive material.[143] Single strand DNA on the device surface attaches to biotinylated complementary strand which, in turn, interacts with magnetically labeled streptavidin. The change in magnetic fringing filed is detected with the sensor. Analogous strategies have been used in immunochemical detection *via* magnetically labeled particles.[144]

Surface plasmon resonance, described above, has been employed successfully together with electrochemistry.[145] The necessary gold layer for supporting plasmons can also be utilized as a conventional electrode. This combination is particularly useful for studying the optical properties of an electropolymerization process under an electric field.[146] In an analogous approach, electrochemical impedance spectroscopy has proven to be extremely useful for monitoring SPR responses for correlation with various electrochemical signals obtained from DNA hybridization.[147] Other optical combination methodologies include wave guide and ellipsometric physics. The former is concerned with evanescent field optical effects combined with electrochemistry of a material surface.[148] Ellipsometry, a well-known method for measuring the thickness of layers and films in electronic fabrication processes, in the combination with electrochemistry has been employed for detecting aspects of protein adsorption on surfaces.[149]

Finally, we mention the use of method combinations with scanning probe technology in the detection of biochemical events. (More detail regarding these techniques for the direct study neural cells is considered in Chapter 5.) The well-known atomic force microscope (AFM) concerns the 'dragging' of a tip across a surface with measurement of the force involved with a cantilever device. It has been used extensively in conjunction with ultra-microelectrode technology where forms of amperometric or potentiometric electrochemistry complement the structural information produced by the AFM tip.[150] This technique is more often than not referred to as scanning electrochemical microscopy and the main application focus has been on surface characterization rather than biochemical binding events.

The scanning Kelvin nanoprobe (SKN) can be considered as an adjunct-type instrument since it measures, simultaneously, the topography and electrical contact potential difference on a substrate. In essence, a vibrating, guarded tip is scanned over a surface and the resulting Kelvin physics is used to detect biochemical binding events such as oligonucleotide duplex formation and immunochemical interactions through alteration of surface dipole charge changes.[151] The instrument has been used successfully to examine cells including neurons on surfaces (see Chapter 6).

References

1. A. Kobmakov, Y. Zhng, G. Chen and M. Moscovits, *Adv. Mater.*, 2003, **15**, 997.
2. E. N. Dattoli, A. V. Davydov and K. D. Benkstein, *Nanoscale*, 2012, **4**, 1760.
3. P. Baldi and G. W. Hatfield, *DNA Microarrays and Gene Expression*, Cambridge University Press, Cambridge, UK, 2002.
4. H. Kaiser, *Pure Appl. Chem.*, 1973, **34**, 35.
5. P. K. Yuen, N. H. Fontaine, M. A. Quesada, P. Mazumder, R. Bergman and E. J. Mozdy, *Lab-on-Chip*, 2005, **5**, 959.
6. L. C. Clark, *Ann. NY Acad. Sci.*, 1962, **102**, 29.
7. L. C. Clark, R. Wolf, D. Granger and Z. Taylor, *Appl. Physiol.*, 1953, **6**, 189.
8. S. J. Updike and J. P Hicks, *Nature*, 1967, **214**, 986.
9. J. Wang, *Electroanalysis*, 2001, **13**, 893.
10. J. Wang, *Chem. Rev.*, 2008, **108**, 814.
11. V. Dugas, A. Elaissari, and Y. Chevalier, in *Recognition Receptors in Biosensors*, ed. M. Zourob, Springer Science + Business Media, New York and London, 2010, Chapter 2, p. 47.
12. D. C. Kim and D. J. Kang, *Sensors*, 2008, **8**, 6605.
13. R. D. Tilton, C. R. Robertson and A. P. Gast, *Langmuir*, 1991, **7**, 2710.
14. L. M. Furtado, H. Su and M. Thompson, *Anal. Chem.*, 1999, **71**, 1167.
15. H. M. Green, *Adv. Protein Chem.*, 1975, **29**, 85.
16. E. A. Bayer and M. Wilchek, *Trends Biochem. Sci.*, 1978, **3**, N237.
17. E. A. Bayer and M. Wilchek (eds.), *Avidin-Biotin Technology*, Vol. 184, Academic Press, San Diego, CA, 1990.
18. G. P. Anderson, M. A. Jacoby, F. S. Ligler and K. D. King, *Biosens. Bioelectron.*, 1997, **12**, 329.
19. R. Gupta and N. K. Chaudhury, *Biosens Bioelectr.*, 2007, **22**, 2387.
20. W. Jin and J. D. Brennan, *Anal. Chim. Acta*, 2002, **461**, 1.
21. R. A. Williams and H. W. Blanch, *Biosens. Bioelectron.*, 1994, **9**, 159.
22. K. Jans, K. Bonroy, G. Reekmans, R. De Parma, S. Peeters, H. Jans, T. Stakenborg, F. Frederix and W. Laureyn, in ed. M.-I. Baraton, *Sensors for Environment, Health and Security*, 2009, p. 277.
23. K. B. Blodgett, *J. Am. Chem. Soc.*, 1935, **57**, 1007.
24. A. Ulman, *An Introduction to Ultrathin Organic Films: From Langmuir-Blodgett to Self-Assembly*, Academic Press, Boston, MA, 1991.
25. A. Zhang, Y. Hou, N. Jaffrezic-Renault, J. Wan, A. Soldatkin and J.-M. Chovelon, *Bioelectrochemistry*, 2002, **56**, 167.
26. Z. Matharu, G. Sumana, S. K. Arya, S. P. Singh, V. Gupta and B. D. Malhotra, *Langmuir*, 2007, **23**, 13188.
27. L. H. Dubois and R. G. Nuzzo, *Ann. Rev. Phys. Chem.*, 1992, **43**, 437.
28. F. Schreiber, *Prog. Surf. Sci.*, 2000, **65**, 151.
29. J. C. Love, L. A. Estroff, J. K. Kriebel, R. G. Nuzzo and G. M. Whitesides, *Chem. Rev.*, 2005, **105**, 1103.

30. C. Vericat, M. E. Vela, G. Benitez, P. Carro and R. C. Salvarezza, *Chem. Soc. Rev.*, 2010, **39**, 1085.
31. H. Hakkinen, *Nature Chem.*, 2012, **4**, 443.
32. S. Sheikh, J. C.-C. Sheng, C. Blaszykowski and M. Thompson, *Chen. Sci.*, 2010, **1**, 271.
33. S. Sheikh, C. Blaszykowski and M. Thompson, *Chem. Commun.*, 2012, **48**, 1305.
34. K. Haupt and K. Mosbach, *Chem. Rev.*, 2000, **100**, 2495.
35. G. Vlatakis, L. I. Anderson, R. Muller and K. Mosback, *Nature*, 1993, **361**, 645.
36. E. L. Holthoff and F. V. Bright, *Acc. Chem. Res.*, 2007, **40**, 756.
37. M. Pohanka and P. Skládal, *J. Appl. Biomed.*, 2008, **6**, 57.
38. D. Grieshaber, R. MacKenzie, J. Vörös and E. Reimhult, *Sensors*, 2008, **8**, 1400.
39. N. J. Ronkainen, H. B. Halsall and W. B. Heineman, *Chem. Soc. Rev.*, 2010, **39**, 1747.
40. J. Wang, *Biosens. Bioelectron.*, 2006, **21**, 1887.
41. J. Korita and K. Stulik, *Ion Selective Electrodes*, 2nd ed., Cambridge University Press, Cambridge, UK, 1983.
42. C. H. Fry and S. E. N. Langley, *Ion-Selective electrodes for Biological Systems*, CRC Press, Boca Raton, FL, 2002.
43. P. Bergveld, *Biosensors*, 1986, **2**, 15.
44. P. Bergveld, *Sens. Actuators, B*, 2003, **88**, 1.
45. M. Yuqing, G. Jianquo and C. Jianrong, *Biotechnol. Adv.*, 2003, **21**, 527.
46. M. J. Schöning and A. Poghossian, *Analyst*, 2002, **127**, 1137.
47. J. Wang, *Electroanalysis*, 2001, **13**, 983.
48. J. Wang, *Chem. Rev.*, 2008, **108**, 814.
49. J. Wang, *Perspect. Bioanal.*, 2005, **1**, 175.
50. Z. Fang, L. Soleymani, G. Pampalakis, M. Yoshimoto, J. A. Squire, E. H. Sargent and S. O. Kelley, *ACS Nano*, 2009, **3**, 3207.
51. J. O. Schenk, E. Meller, M. E. Rice and R. N. Adams, *Brain Res.*, 1983, **277**, 1.
52. A. B. Kharitonov, J. Wasserman, H. Katz and H. Willner, *J. Phys. Chem., B*, 2001, **105**, 4205.
53. R. De Marco, A. Ng and D. Panduwinata, *Electroanalysis*, 2008, **20**, 313.
54. L. Alfonta, A. Bardea, O. Khersonsky and I. Willner, *Biosens. Bioelectron.*, 2001, **16**, 675.
55. A. Bratov, J. Ramón-Azcón, N. Abramova, A. Merlos, J. Adrian, F. Sánchez-Baeza, M.-P. Marco and C. Dominguez, *Biosens. Bioelectron.*, 2008, **24**, 729.
56. S. Pan and L. Rothberg, *Langmuir*, 2005, **21**, 1022.
57. A. K. Wanekaya, W. Chen, N. V. Myung and A. Mulchandani, *Electroanalysis*, 2006, **18**, 533.
58. L. Patolsky, G. Zheng and C. M. Lieber, *Anal. Chem.*, 2006, **78**, 4260.
59. U. Yogeswaran and S.-M. Chen, *Sensors*, 2008, **8**, 280.

60. O. Knopfmacher, A. Tarasov, W. Fu, M. Wipf, B. Niesen, M. Calame and C. Schönenberger, *Nano Lett.*, 2010, **10**, 2268.
61. K.-I. Chen, B.-T. Li and Y.-T. Chen, *NanoToday*, 2011, **6**, 131.
62. J. E. Roederer and G. J. Bastians, *Anal. Chem.*, 1983, **55**, 2333.
63. G. Calabrese, H. Wohltjen and M. K. Roy, *Anal. Chem.*, 1875, **59**, 833.
64. P. L. Kpnash and G. J. Bastiaans, *Anal. Chem.*, 1980, **52**, 1929.
65. W. G. Cady, *Piezoelectricity*, Dover Publications, New York, Volume 1, 1964, p. 4.
66. Lord Kelvin, *Phil. Mag.*, 1983, **36**, pages 331, 342, 384 and 453.
67. Lord Rayleigh, *Proc. London Math. Soc.*, 1885, **17**, 4.
68. M. Sklodoswska-Curie, *Research on Radioactive Substances*, Sorbonne, Paris, 1903.
69. C. Z. Rosen, B. V. Hiramath and R. Nwnham (eds.), *Piezoelectricity*, American Institute of Physics, New York, 1992.
70. D. S. Ballantine, R. M. White, S. J. Maartin, A. J. Ricco, E. T. Zellers, G. C. Frye and H. Wohltjen, *Acoustic Wave Sensors: Theory, Design and Physico-Chemical Applications*, Academic Press, San Diego and London, 1996, p. 25.
71. G. Sauerbrey, *Z. Phys.*, 1959, **155**, 206.
72. E.-L. E. Lyle, *Acoustic Coupling of Transverse Waves as a Mechanism for the Label-Free Detection of Biomolecular Interactions*, PhD thesis, University of Toronto, 2004.
73. K. K. Kanazawa and J. G. Gordon, *Anal. Chim. Acta*, 1985, **175**, 99.
74. J. S Ellis and M. Thompson, *Phys. Chem. Chem. Phys.*, 2004, **6**, 4928.
75. J. S. Ellis and M. Thompson, *Chem. Commun.*, 2004, Number 11, 1310.
76. J. S. Ellis and M. Thompson, *Langmuir*, 2010, **26**, 11558.
77. M. Edvardsson, M. Rodahl, B. Kasemo and F. Hook, *Anal Chem.*, 2005, **77**, 4918.
78. X. Wang, J. S. Ellis, P. Sundaram, E.-L. Moore and M. Thompson, *Mol. Biosys.*, 2006, **2**, 184.
79. D. H. Ather and V. Relpa, *J. Biophys. Chem.*, 2012, **3**, 211.
80. H. Su, K. M. R. Kallury, M. Thompson and A. Roach, *Anal. Chem.*, 1994, **66**, 769.
81. X. Wang, J. S. Ellis, C.-D. Kan, R.-K. Li and M. Thompson, *Analyst*, 2008, **133**, 85.
82. M. Thompson, S. M. Ballantyne, A. C. Stevenson and C. R. Lowe, *Analyst*, 2003, **128**, 1048.
83. S. Sheikh, C. Blaszykowski and M. Thompson, *Talanta*, 2011, **85**, 816.
84. M. Thompson and D. C. Stone, *Surface-Launched Acoustic Wave Sensors: Chemical Sensing and Thin Film Characterization*, Wiley-Interscience, New York and Chichester, 1997.
85. F. Josse, F. Bender and R. W Cernosek, *Anal. Chem.*, 2001, **73**, 5937.
86. M. Tom-Moy, R. L. Baer, D. Spra-Solomon and T. P. Doherty, *Anal. Chem.*, 1995, **67**, 1510.
87. K. Saha, F. Bender and E. Gizeli, *Anal. Chem.*, 2003, **75**, 834.

88. K. Lange, B. R. Rapp and M. Rapp, *Anal. Bioanal. Chem.*, 2008, **391**, 1509.
89. A. C. Stevenson and C. R. Lowe, *Sens. Actuators, A*, 1999, **72**, 32.
90. P. C. H. Li and M. Thompson, *Anal. Chem.*, 1996, **68**, 2590.
91. X. Fan, I. M. White, S. J. Shopova, H. Zhu, J. D. Suter and Y. Sun, *Anal. Chim. Acta*, 2000, **620**, 8.
92. K. Kneipp, R. Aroca, H. Kneipp and E. Wentrup-Byrne, *New Approaches in Biomedical Spectroscopy*, ACS Symposium Series 963, American Chemical Society, Washington DC, 2007.
93. *Analyst*, 2009, **134**, themed issue on clinical diagnostics.
94. N. Stone, C. Kendall, N. Shepherd, P. Crow and H. Barr, *J. Raman Spectrosc.*, 2002, **33**, 564.
95. M. Almond, J. Hutchings, C. Kendall, J. C. C. Day, O. A. C. Stevens, G. R. Lloyd, N. A. Shepherd, H. Barr and N. Stone, *J. Biomed. Optics*, 2012, **17**, 081421.
96. L.-L. Tay, R. G. Tremblay, J. Hulse, B. Zurakowski, M. Thompson and M. Bani-Yaghoub, *Analyst*, 2011, **136**, 1620.
97. B. T. Cunningham, in *Label-Free Biosensors*, ed. M. A. Cooper, Cambridge University Press, Cambridge, UK.
98. C. I. L. Justino, T. A. Rocha-Santos and A. C. Duarte, *Trends Anal. Chem.*, 2010, **29**, 1172.
99. E. A. H. Hall, *Biosensors*, Biotechnology Series, Open University Press, Buckingham, UK, 1990, p. 141.
100. R. A Wolthuis, D. McCrae, J. C. Hartl, E. Saaski, G. L Mitchell, K. Garcin and R. Willard, *IEEE Trans. Biomed. Eng.*, 1992, **39**, 185.
101. Y. Li, Y. Fan, L. Zhang, S. Ma and J. Zheng, *Proc. SPIE*, 2000, **4077**, 242.
102. J. P. Golden, G. P. Anderson and F. S. Ligelr, *IEEE Trans, Biomed. Deng.*, 1994, **41**, 585.
103. C. R. Taitt, G. P. Anderson and F. S. Ligler, *Biosens. Bioelectron.*, 2005, **20**, 2470.
104. S. Balr and Y. Chen, *Appl. Optics*, 2001, **40**, 570.
105. P. A. E. Piunno, U. J. Krull, R. H. E. Hudson, M. J. Damha and H. Cohen, *Anal. Chim. Acta*, 1994, **288**, 206.
106. P. A. E. Piunno, U. J. Krull, R. H. E. Hudson, M. J. Damha and H. Cohen, *Anal. Chem.*, 1995, **67**, 1635.
107. A. M. Valasex, C. A. Lana, S. Tu, M. T. Morgan and A. S. K. Bhunia, *Sensors*, 2009, **9**, 5819.
108. M. E. Bosch, A. J. R. Sánchez, F. S. Rojas and C. B. Ojeda, *Sensors*, 2007, **7**, 797.
109. J. Homola (ed.), *Surface Plasmon Resonance Based Sensors*, Springer, Berlin, 2006.
110. J. Homola, S. S. Yee and G. Gauglitz, *Sens. Actuators, B*, 1999, **4**, 3.
111. X. D. Hoa, A. G. Kirk and M. Tabrizian, *Biosens. Bioelectron.*, 2007, **3**, 151.
112. R. L. Rich and D. G. Miszka, *Curr. Opin. Biotech.*, 2000, **11**, 54.
113. I. Abdulhalim, M. Zourob and A. Lakhyakis, *Electromagnetics*, 2008, **28**, 215.

114. X. Guo, *J. Biophotonics*, 2012, **5**, 483.
115. J. Homola, *Anal. Bioanal. Chem.*, 2003, **377**, 528.
116. E. Kretschmann, *Z. Phys.*, 1971, **241**, 313.
117. L. He, M. D. Musick, S. R. Nicewarner, F. G. Salinas, S. J. Benkovic, M. J. Natan and C. D. Keating, *J. Am. Chem. Soc.*, 2000, **122**, 9071.
118. I. H. El-Sayed, Xi. Huang and M. A. El-Sayed, *Nano Lett.*, 2005, **5**, 829.
119. G. J. Wegner, H. J. Lee and R. M. Corn, *Anal. Chem.*, 2002, **74**, 5161.
120. P. Safsten, *Methods Mol. Biol.*, 2009, **524**, 67.
121. V. Chabot, C. M. Cuerrier, E. Escher, V. Aimez, M. Grandbois and P. G. Charette, *Biosens. Bioelectron.*, 2009, **15**, 1667.
122. J. G. Quinn, S. O'Neill, S. Doyle, C. M. McAtamney, D. Diamond, B. D. MacCraith and R. O'Kenndy, *Anal. Biochem.*, 200, **281**, 135.
123. M. Skorobogatiy and A. V. Kabashin, *Applied. Phys. Lett.*, 2006, **89**, 143518.
124. B. P. Nelson, T. E. Grimsrud, M. R. Liles, R. M. Goodman and R. M. Corn, *Anal. Chem.*, 2001, **73**, 1.
125. S. Scarano, M. Mascini, A. P. F. Turner and M. Minunni, *Biosens. Bioelectron.*, 2010, **26**, 957.
126. S. Shiekh, C. Blaszykowski and M. Thompson, *Anal. Lett.*, 2008, **41**, 2525.
127. R. G. Heideman, R. P. H. Kooyman and J. Greve, *Sens. Actuators, B*, 1993, **10**, 209.
128. R. G. Heideman and P. V. Lambeck, *Sens. Actuators, B*, 1999, **61**, 100.
129. A. Brandenburg, *Sens. Actuators, B*, 1997, **39**, 266.
130. B. H. Schneider, E. L. Dickinson, M. D. Vach, J. V. Hoijer and L. V. Howard, *Biosens. Bioelectron.*, 2000, **15**, 13.
131. M. Zhao, D. Nolte, W. Cho, F. Regnier, M. Varma, G. Lawrence and J. Pasqua, *Clin. Chem.*, 2006, **52**, 2135.
132. A. P. Alivisatos, *ACS Nano*, 2008, **2**, 1514.
133. Y. Masumoto and T. Takagahara (eds.), *Semiconductor Quantum Dots*, Springer-Verlag, Berlin, 2002.
134. I. L. Medintz and H. Mattoussi, *Phys. Chem. Chem. Phys.*, 1009, **11**, 17.
135. J. B. Delehantym, I. L Medintz and H. Mattoussi, *Anal. Bioanal. Chem.*, 2009, **393**, 1091.
136. A. M. Smith, X. Gao and S. Nie, *Photchem. Photobiol.*, 2004, **80**, 377.
137. W. R. Algar, A. J. Tavares and U. J. Krull, *Anal. Chim. Acta*, 2010, **673**, 1.
138. J. B. Jackson and N. J. Hals, *Proc. Natl. Acad. Sci. U. S. A.*, 2004, **101**, 17930.
139. M. W. Meyer and E. A. Smith, *Analyst*, 2011, **136**, 3542.
140. X. Zhang, C. R. Yonzon, M. A. Young, D. A. Stuart and R. P. Van Duyne, *IEEE Proc. Nanobiotechnol.*, 2005, **152**, 195.
141. O. Oehler, M. Seifert, S. Cliffe and K. Mosbach, *Infrared Phys.*, 1985, **25**, 319.
142. A. Mukherjee, I. Dunayevskiy, R. Go, A. Tsekoum, X. Wang, J. Fam and C. K. N. Patel, *Applied Optics*, 2008, **47**, 1543.

143. D. L. Graham, H. A. Ferreira and P. P. Freitas, *Trends Biotechnol.*, 2004, **2**, 455.
144. J Richardson, P. Hawkins and R. Luxton, *Biosens. Bioelectron.*, 2001, **16**, 989.
145. J. Xiang, J. Guo and F. Zhou, *Anal. Chem.*, 2006, **78**, 1418.
146. X. Kang, Y. Jin, G. Chang and S. Dong, *Langmuir*, 2002, **18**, 1713.
147. M. Stambouli, V. Boukherroub, R. Szunerits and S. Szunerits, *Analyst*, 2008, **133**, 1097.
148. J. P. Bearinger, J. Vörös, J. A. Hubell and M. Textor, *Biotechnol. Bioeng.*, 2003, **82**, 465.
149. P. Ying, A. S. Viana, L. M. Abrantes and G. Jin, *J. Colloid Interface Sci.*, 2004, **279**, 95.
150. A. Anne, E. Cambril, A. Chovin, C. Demaille and C. Goyer, *ACS Nano*, 2009, **3**, 2927.
151. L.-E. Cheran, S. Jounstone, S. Sadeghi and M. Thompson, *Meas. Sci. Technol.*, 2007, **18**, 567.

Selected Bibliography of Biosensor Technology 1987–2012

H. Baltes, J. Hesse and J. G. Korvink (eds.), *Sensors Update Vol. 9: Sensors Technology, Applications, Markets*, Wiley-VCH, Weinheim, 2001.

H. Baltes, G. K. Fedder and J. G. Korvink (eds.), *Sensors Update Vol. 10: Sensors Techonology, Applications, Markets*, Wiley-VCH, Weinheim, 2002.

D. Barcelo and P.-D. Hansen (eds.), *The Handbook of Environmental Chemistry Vol. 5-J: Biosensors for the Environmental Monitoring of Aquatic Systems: Bioanalytical and Chemical Methods for Endocrine Disruptors*, Springer-Verlag, Berlin and Heidelberg, 2009.

H. H. Bau, N. F. de Rooij, B. Kloeck (eds.), *Sensors: A Comprehensive Survey Vol. 7: Mechanical Sensors*, Wiley-VCH, Weinheim, 1995.

U. Bilitewski and A. Turner (eds.), *Biosensors for Environmental Monitoring*, Harwood Academic, Amsterdam, 2000.

A. E. G. Cass, F. S. Ligler and B. D. Hames (eds.), *Immobilized Biomolecules in Analysis: A Practical Approach*, Oxford University Press, Oxford, 1998.

R. Comeaux and P. Novotny (eds.), *Biosensors: Properties, Materials and Applications*, Nova Science Publishers, Hauppauge, NY, 2010.

J. Cooper and A. E. G. Cass (eds.), *Biosensors*, 2nd ed., Oxford University Press, Oxford, 2003.

M. A. Cooper (ed.), *Label-free Biosensors: Techniques and Applications*, Cambridge University Press, Cambridge, UK, 2009.

G. Costa and S. Miertus (eds.), *Trends in Electrochemical Biosensors: Proceedings of the Conference*, World Scientific Publishing, Singapore, 1992.

A. J. Cunningham, *Introduction to Bioanalytical Sensors*, John Wiley & Sons, Chichester, 1998.

A. B. Dahlin, *Advances in Biomedical Spectroscopy Vol. 4: Plasmonic Biosensors: An Integrated View of Refractometric Detection*, IOS Press, Lansdale, PA, 2012.
B. R. Eggins, *Biosensors: An Introduction*, John Wiley & Sons, Chichester, 1997.
B. R. Eggins, *Chemical Sensors and Biosensors*, John Wiley & Sons, Chichester, 2002.
P. Fabry and J. Fouletier (eds.), *Chemical and Biological Microsensors: Applications in Fluid Media, ISTE Ltd, London,* 2010.
K. R. Fox and T. Brown (eds.), *DNA Conjugates and Sensors*, Royal Society of Chemistry, Cambridge, UK, 2012.
D. M. Fraser (ed.), *Biosensors in the Body: Continuous In Vivo Monitoring*, John Wiley & Sons, Chichester, 1997.
R. Freitag (ed.), *Biosensors in Analytical Biotechnology*, Academic Press, Austin, TX, 1996.
E. Gizeli and C. R. Lowe (eds.), *Biomolecular Sensors*, Taylor and Francis, London2002.
L. Gorton and D. Barcelo (eds.), *Comprehensive Analytical Chemistry Vol. 44: Biosensors and Modern Biospecific Analytical Techniques*, Elsevier, Amsterdam, 2005.
M. Grattarola and G. Massobrio, *Bioelectronics Handbook*, McGraw-Hill, New York, 1998.
E. A. H. Hall, *Biosensors*, Prentice Hall, Englewood CLiffs, NJ, 1991.
R. V. Harrison (ed.), *Chemical Sensors: Properties, Performance and Applications*, Nova Science Publishers, Hauppauge, NY, 2010.
P. J. Hesketh (ed.), *BioNanoFluidic MEMS*, Springer Science + Business Media, New York and London, 2008.
J. D. Higgins, G. Di Giovanni and V. J. Harwood, *Overcoming Molecular Sample Processing Limitations: New Platform Technologies, RNA and DNA Extraction Strategies and Fibre-Optic Biosensors*, IWA Publishing, London, 2003.
C. P. Hollenberg and H. Sahm (eds.), *Biotec 2: Biosensors and Environmental Biotechnology*, Gustav Fischer, Stuttgart, 1988.
T. Jacobs, *Miniaturized Thermal Flow and Impedimetric Sensors for the Inline Chemical Process Analysis in Micro-plants*, Shaker Verlag, Aachen, Germany, 2011.
D. W. Jeffrey and B. Madden (eds.), *Bioindicators and Environmental Management*, Academic Press, London, 1991.
R. M. Joshi (ed.), *Biosensors*, Isha Books, Delhi, 2006.
G. K. Knopf and A. S. Bassi (eds.), *Smart Biosensor Technology*, CRC Press, Boca Raton, FL, 2007.
C. Kumar (ed.), *Nanomaterials for Biosensors*, Wiley-VCH, Wienheim, 2007.
M. Lambrechts and W. Sansen, *Biosensors: Microelectrochemical Devices*, Institute of Physics Publishing, Bristol, 1992.
Z. Liron, A. Bromberg and M. Fisher (eds.), *Novel Approaches in Biosensors and Rapid Diagnostic Assays*, Plenum Publishers, New York, 2001.

B. D. Malhotra and A. Turner (eds.), *Advances in Biosensors Vol. 5: Perspectives in Biosensors*, Elsevier Science, Amsterdam, 2003.

R. S. Marks, D. C. Cullen, I. Karube, C. R. Lowe and H. H. Weetall (eds.), *Handbook of Biosensors and Biochips*, Wiley-Blackwell, Oxford, 2007.

M. Mascini and I. Palchetti (eds.), *Nucleic Acid Biosensors for Environmental Pollution Monitoring*, Royal Society of Chemistry, Cambridge, UK, 2011.

A. Merkoci (ed.), *Biosensing Using Nanomaterials*, John Wiley & Sons, Chichester, 2009.

P. Millner (ed.), *Biosensors*, Scion Publishing Ltd., 2009.

V. M. Mirsky and O. S. Wolfbeis (eds.), *Springer Series on Chemical Sensors and Biosensors Vol. 2: Ultrathin Electrochemical Chemo- and Biosensors: Technology and Performance*, Springer-Verlag, Berlin and Heidelberg, 2004.

P. T. Moseley and J. Crocker, *Sensor Materials*, Institute of Physics Publishing, Bristol, 1996.

A. Mulchandani and K. R. Rogers (eds.), *Methods in Biotechnology Vol. 6: Enzyme and Microbial Biosensors: Techniques and Protocols*, Humana Press, Clifton, NJ, 1998.

A. Mulchandani and O. A. Sadik (eds.), *ACS Symposium Series Vol. 762: Chemical and Biological Sensors for Environmental Monitoring*, American Chemical Society, Washington DC, 2000.

D. Nikolelis (ed.), *NATO Science for Peace and Security Series A: Chemistry and Biology: Portable Chemical Sensors*, Springer Science + Business Media, New York and London, 2012.

D. Nikolelis, U. J. Krull, J. Wang and M. Mascini (eds.), *NATO ASI Series 2: Environment Vol. 38: Biosensors for Direct Monitoring of Environmental Pollutants in Field*, Kluwer Academic Publishers, Dordrecht, 1997.

R. J. M. Palma, D. L. Allara and M. Pishko, *Nanostructures for Biosensing: From Sensing Principles to Applications*, Royal Society of Chemistry, Cambridge, UK, 2012.

A. Rasooly and K. E. Herold (eds.), *Biosensors and Biodetection: Methods and Protocols Vol. 1: Optical Based Detectors*, Humana Press, Clifton, NJ, 2009.

K. R. Rogers and A. Mulchandani (eds.), *Methods in Biotechnology Vol. 7: Affinity Biosensors: Techniques and Protocols*, Humana Press, Clifton, NJ, 1998.

K. R. Rogers, A. Mulchandani and W. Zhou (eds.), *ACS Symposium Series Vol. 613: Biosensor and Chemical Sensor Technology: Process Monitoring and Control*, American Chemical Society, Washington DC, 1995.

A. Sadana, *Engineering Biosensors: Kinetics and Design*, Academic Press, Dordrecht, 2002.

A. Sadana, *Biosensors: Kinetics of Binding and Dissociation using Fractals*, Elsevier Science, Amstersdam, 2003.

A. Sadana, *Binding and Dissociation Kinetics for Different Biosensor Applications using Fractals*, Elsevier, Amsterdam, 2005.

F. Scheller and R. D. Schmid (eds.), *GBF Monographs Vol. 17: Biosensors: Fundamentals, Technologies and Applications*, Wiley-VCH, Weinheim, 1992.

R. D. Schmid, G. G. Guilbault, I. Karube, H. L. Schmid and L. B. Wingard (eds.), *GBF Monographs Vol. 10: Biosensors International Workshop 1987 Proceedings*, Wiley-VCH, Weinheim, 1987.

R. D. Schmid and F. Scheller (eds.), *GBF Monographs Vol. 13: Biosensors: Applications in Medicine, Environmental Protection and Process Control*, Wiley-VCH, Weinheim, 1989.

J. Schultz, M. Mrksich, S. N. Bhatia, D. J. Brady, A. J. Ricco, D. R. Walt, C. L. Wilkins (eds.), *Biosensing: International Research and Development*, Springer-Verlag, New York, 2006.

A. O. Scott (ed.), *Biosensors for Food Analysis*, Woodhead Publishing, Cambridge, UK, 1998.

P. Siciliano (ed.), *Sensors for Environment Control: Proceedings of the International Workshop or New Developments*, World Scientific Publishing, Singapore, 2003.

B. Sonnleitner and T. Scheper (eds.), *Advances in Biochemical Engineering Biotechnology Vol. 66: Bioanalysis and Biosensors for Bioprocess Monitoring*, Springer-Verlag, Berlin and Heidelberg, 2000.

U. E. Spichiger-Keller, *Chemical Sensors and Biosensors for Medical and Biological Applications*, Wiley-VCH, Weinheim, 1998.

R. B. Thompson, *Fluorescence Sensors and Biosensors*, Taylor and Francis, London, 2006.

A. Turner (ed.), *Advances in Biosensors: Supplement Vol. 1: Chemical Sensors for in Vivo Monitoring*, JAI Press, Greenwich, CT, 1993.

A. Turner and R. Renneberg (eds.), *Advances in Biosensors Vol. 4: Biosensors: A Chinese Perspective*, JAI Press, Greenwich, CT, 1999.

A. Turner and Y. M. Yevdokimov (eds.), *Advances in Biosensors Vol. 3: Biosensors: A Russian Perspective*, JAI Press, Greenwich, CT, 1995.

A. Turner, I. Karube and G. S. Wilson (eds.), *Biosensors: Fundamentals and Applications*, Oxford University Press, Oxford, 1987.

Y. Umasankar, S. A. Kumar and S.-M. Chen (eds.), *Nanostructured Materials for Electrochemical Biosensors*, Nova Science Publishers, Hauppauge, NY, 2009.

M. Valcarcel and M. D. L. De Castro, *Flow-through (Bio)Chemical Sensors*, Elsevier Science, Amsterdam, 1994.

V. C. Yang and T. T. Ngo (eds.), *Biosensors and their Applications*, Plenum Publishing, New York, 2000.

P. Wang and Q. Liu (eds.), *Cell-Based Biosensors: Principles and Applications*, Artech House, Norwood, MA, 2010.

P. Wide, *Artificial Human Sensors: Science and Applications*, Pan Stanford Publishing, Singapore, 2012.

D. L. Wise and L. B. Wingard, Jr. (eds.), *Biosensors with Fiberoptics*, Humana Press, Clifton, NJ, 1991.

M. Zourob (ed.), *Recognition Receptors in Biosensors*, Springer Science+Business Media, New York and London, 2010.

M. Zourob, A. Lakhtakia and G. Urban (eds.), *Springer Series on Chemical Sensors and Biosensors Vol. 7: Optical Guided-wave Chemical and Biosensors I*, Springer-Verlag, Berlin and Heidelberg, 2010.

CHAPTER 2
The Cell-Substrate Surface Interaction

2.1 Cells and Surfaces

Any treatment of the application of sensors, and devices in general, in neuroscience must take account of how cells behave at the surface of solids. This is the case whether studies are of an *in vitro* nature or whether implantables are involved. The key issues are how surfaces influence the response of cells and just as important, as it pertains to implantable technology, what do cells and tissues do to the device. A host of factors are anticipated to be important such as surface free energy, morphology, crystal structure, role of interfacial water and ionic charge, and the nature of chemical functional groups present on the substrate (device) surface. A number of these properties are inextricably integral to each other, for example, surface free energy and both functional group chemistry and morphology.

How the various physicochemical factors influence cells will obviously depend on the structure and components that define the biological moiety. Essentially, eukaryotic cells at the simplest level consist of a lipid bilayer membrane which encloses the cytoplasm and internal cellular machinery. They are embedded in a material termed the extracellular matrix (ECM), which serves as a scaffold for the support of cellular populations. Over recent years it has become abundantly clear that interaction of surfaces with the various cellular and extracellular components can affect a large number of parameters, including not only biochemical behavior, but also mechanical properties. In this section we describe, at a concise level, our present understanding of how solid surfaces influence the behavior of cells. For the non-chemistry specialist we start with key surface chemical aspects and for the non-biology specialist

2.2 Substrate Surface Parameters: A Précis

Biological cells of any type will exhibit a plethora of 'exterior' functional groups, originating from protein, lipid and saccharide moieties, together with domains of polar and hydrophobic character orchestrated in a highly dynamical fashion. Accordingly, with respect to their interaction with bare or coated substrates, the existence of interfacial functional groups and their spatial distribution, roughness or morphology, physical pattern and structure (pillars, *etc.*), free energy, charge and elasticity are all expected to play pivotal and concerted roles.

As would be expected surface scientists have employed a large number of analysis techniques for the characterization of surfaces and many of these have been employed in connection with biomaterials and sensor technology. Indeed there is much common ground with respect to surface analysis of devices and materials. Included among this armamentarium are X-ray photoelectron spectroscopy, atomic force microscopy, scanning tunneling microscopy, secondary ion mass spectrometry, electron microscopy, confocal fluorescence microscopy and contact angle measurement.[1,2] The last of these is important in terms of the study of the surface free energy of a bare or adlayer-treated surface. This leads to an appraisal of the critical role of interracial thermodynamics with regard to the cell–substrate interaction, at least at the level of a first approximation. This implies that such an interaction can take place if there is an overall reduction in interfacial fee energy, with the caveat that surface morphology will undoubtedly play a key role.

There are a huge number of electronegative atoms such as such as O and N present on the exterior 'boundary' of cells which originate from membrane proteins and lipidic moieties and other biochemical sources. Accordingly, this leads to the possibility for extensive intermolecular hydrogen bonding with a polar or hydrophilic surface. This involves an essentially electrostatic interaction of some $2-10\,\text{kcal}\,\text{mol}^{-1}$ where functional groups with the hydrogen atom may act as a donor or acceptor for formation of the bond. Since nonpolar domains are also expected on the surface of a cell the hydrophobic effect, about which much has been written, will also play a role. This effect is generally considered to originate from entropic considerations associated with the exclusion of water and disruption of water-based hydrogen bonds. For an excellent look at the hydrogen and the water molecule please consult ref. 3.

Given the importance of interfacial free energy, hydrophilic/hydrophobic modification has received great attention with the aim of enhancing cell attachment, or indeed avoidance of cellular interactions for particular applications. A surface can be simply rendered hydrophilic by the introduction of negatively charged groups such as the carboxyl ($-COO^-$) entity. Neutral hydrophilic surfaces can be obtained by hydroxyl or amide groups, and cationic hydrophilic surfaces are made by introduction of different amino groups

(–NH$_2$, –NHR or –NR$_2$). As we shall see later, coating of hydrophilic surfaces with polysaccharides (*e.g.* dextrans) has been recognized as advantageous for neuronal cell adhesion because each monomer within the chain carries up to three hydroxyl groups. As discussed later so are also the so-called biomimetic peptides. A particularly prevalent example is the key tripeptide sequence RGD (Arg-Gly-Asp), which is present in fibronectin.

Aside from the electrostatic hydrogen bond, there are potentially other charge-based interactions that can occur between the exterior of a biological cell and a substrate surface, especially when the latter is expected to be charged as may be the case for electrode materials. These range from formal charge–charge interactions to those involving induced dipoles. A summary compendium of the energy of these forces and their distance dependence is given in Table 2.1. In this respect an interesting and potentially important interaction, which has not been widely discussed, is that of the membrane surface dipole electrical potential with either surface formal charge or dipoles.[4] This potential originates largely from certain lipid components of the biological membrane (the lipid composition of some types of cell is shown in Table 2.2). The dipoles of the headgroups of lipid molecules at the surface are aligned within an approximate (highly dynamic) sheet resulting in the surface potential, which is distinct from the conventional potentials outlined above. Interaction of this potential with, for example, the Gouy–Chapman double layer present on electrodes could constitute an important adhesive, or indeed, repulsive force.

Table 2.1 Basic energy equations for electrostatic interactions.

Interaction	Potential energy (U)
Charge–charge	$U = q_1 (q_2/r)$
Charge–dipole	$U = q \mu \cos \theta / (Dr^2)$
Dipole–dipole	$U = -\mu_1 \mu_2 (\cos \theta_{12} - 3 \cos \theta_1 \cos \theta_2) / 4 \pi \varepsilon_0 r_{12}^3$
Charge–induced dipole	$U = -1/2 \, \alpha \, q^2 / (r^4 D^2)$

Key: q is formal charge, r is distance, μ is dipole moment, D is dielectric constant, ε_0 is permittivity, α is polarizability, and θ is the angle between charge and/or dipole (center). Subscripts 1 and 2 label each of two dipoles.

Table 2.2 Examples of lipid composition of eukaryotic plasma membranes.

Cell	PC	PE	PI	PS	PA	CL	Sph	Ch	CB	SL
Fibroblast	43.2	16.1	7.6	6.4	1.5	–	12.2	13.0	–	–
Schwann	44.0	34.0	9.6	2.8	–	–	29.6	0.6	–	–
Dorsal root ganglia (neuronal cell)	28.9	16.5	5.8	3.1	–	–	2.8	15.4	3.1	1.6
Reticular nucleus	34.0	27.5	–	–	–	6.3	12.6	19.6	–	–
Human T cell	43.0	32.9	–	7.4	–	–	10.4	–	–	–

Key: PC, phosphatidylcholine; PE, phosphatidylethanolamine; PI, phosphatidylinositol; PS, phosphatidylserine; PA, phosphatidic acid; CL, cardiolipin; Sph, sphingomyelin; Ch, cholesterol; CB, cerebrosides; SL, sulfatide.

Of great significance are both the spatial chemical functionality and physical morphology of the substrate surface. Cells appear to recognize the roughness of a surface in terms of response which is very important, for example, as this property pertains to implanted materials such as steel. This observation has led to the orchestrated design of substrate morphology by photolithographic and other techniques, for example, to produce pillars and wells which influence cell adhesion and response. Another factor that has received relatively scant attention is the rheological character, especially elasticity, of the substrate. Experiments in recent years have shown that various cells are capable of 'sensing' the rigidly or elastic properties of the substrate. Both the effects of surface morphology and elasticity on cellular properties are discussed later in the text.

2.3 The Eukaryotic Cell and Environment: A Précis

Animal eukaryotic cells have a dimension of the order of 50 nm and have various organelles enclosed within a plasma membrane composed of a mixture of amphipathic phospholipids and sterols. The membrane incorporates a wide variety of proteins which have several functions including molecular transport and biochemical signaling.[5] An example of an internal organelle is the mitochondrion which is responsible for the production of adenosine triphosphate (ATP).[6] This type of cell also possesses an internal support structure called the 'cytoskeleton', which is composed of protein filaments. These moieties have several functions that include cell motion and muscle contraction. It is important to note that when it comes to the external environment of the cell, the cytoskeleton also plays a crucial role, especially with respect to the property of adherence.

Cells of whatever type are embedded in the ECM, which is composed of fibrous proteins and a polysaccharide gel-like matrix (Figure 2.1). The ECM is

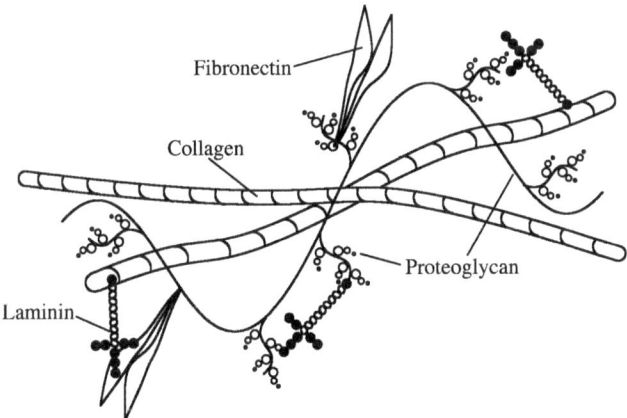

Figure 2.1 Some components of the extracellular matrix.

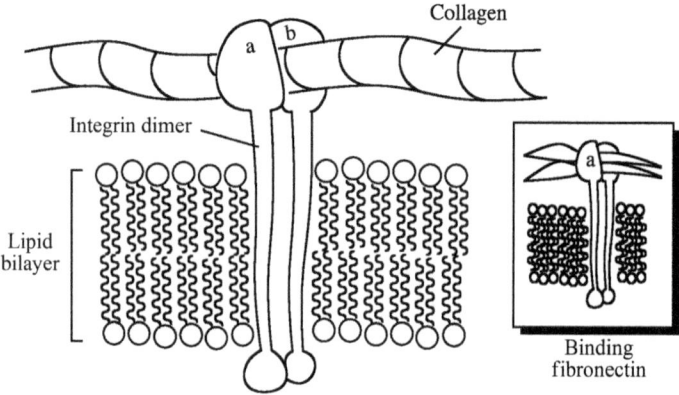

Figure 2.2 Integrin subunits embedded lipid membrane with connection to ECM.

highly variable in nature with a particular composition being associated with a specific function of the tissue involved. The chemistry exhibited by the ECM macromolecules is connected to cellular proliferation, structural integrity and adhesive properties.[7] Examples of the structural proteins are collagen and elastin and for those involved in adherence, fibronectin and laminin. The ECM in connective tissue has incorporated cells relatively sparsely spaced, whereas in the epithelium they are in close proximity with each other. The interaction of various cells with the ECM is mediated by membrane-bound proteinaceous entities known as integrins. These proteins effectively connect together the cytoskeleton with proteins in the ECM such as laminin, fibronectin and collagen. They are present in the lipid membrane in the form of a dimer and take a variety of forms capable of attaching to ECM proteins (Figure 2.2).[8] The interaction between integrin and ECM components is complex and involves conformational changes in the former often described in terms of focal adhesions. This binding event is associated with a conserved RGD sequence in the ECM proteins.[9] As one example of the processes involved we mention the chemistry of fibronectin, which is a fibril-forming glycoprotein with a dimeric structure composed of 230–270 kD monomers. The protein contains domains that bind to a number of moieties such as collagen, ECM proteins and the cell membrane. The cell binding region of the protein possesses the RGD sequence which attaches by hydrogen bonding to glycine residues in integrins.[10]

The ECM chemistry is highly relevant to protocols for the modification of surfaces. Not surprisingly, in order to promote solid–substrate binding of various cells, the interface has first been treated with ECM proteins on a so-called inert background. Such efforts are discussed in more detail later is this chapter.

2.4 The Neuron: A Précis

Contemporary neuroscience is a highly interdisciplinary field incorporating biology, chemistry physics, engineering and computer science among many

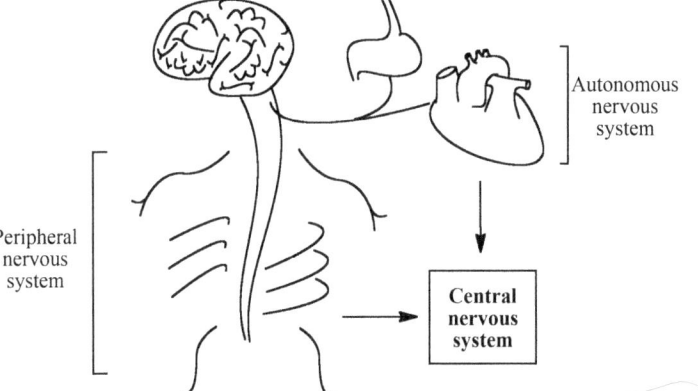

Figure 2.3 Schematic of the nervous sytem.

areas of scientific endeavor.[11] Research conducted in the field ranges all the way from the cellular biology of nerve cells to imaging and brain function. It is considered to be divided into a number of areas which include neurophysiology, cognitive science, and computational, molecular, endocrinological and engineering neuroscience.[12–16] At the heart of these activities is the study of the structure and function of single nerve cells or of the massive collections of neurons in the nervous system.

The human central nervous system (CNS) is composed of three anatomical units which are designed to act in a unified fashion (Figure 2.3). These are the brain and spinal cord, peripheral nervous system and the autonomous nervous system. (For the non-biologist, the structure and functions of the brain are described beautifully in the superb, illustrated text by Rita Carter.[17]) The second of these systems permeates the whole body and consists of a very complex network of nerves that emanate from both cranial and spinal nerves. Information both passes to and from the brain in this system, which incorporates an astonishing range of hierarchical levels ranging across the brain, networks, neurons and, of course, molecular chemistry. We turn to a concise look at the latter two of these components in more detail, since much of the application of devices in neuroscience is connected to this area.

2.4.1 Anatomy and Types

The neuron is an electrically excitable cell that conveys information through chemical and electrical signaling mechanisms. The cells are incorporated into complex networks which display massive interconnectivity. They exhibit a number of functions which include information processing from the outside world, such as the sensory neurons, and motor neurons that are responsible for signaling to muscle tissue.[12] Although neurons are variable in structure and display a wide variety of functions, a representative cell is depicted in

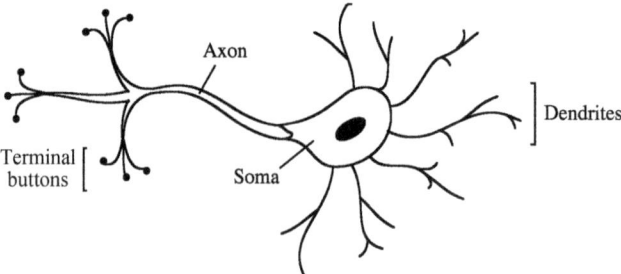

Figure 2.4 The neuron.

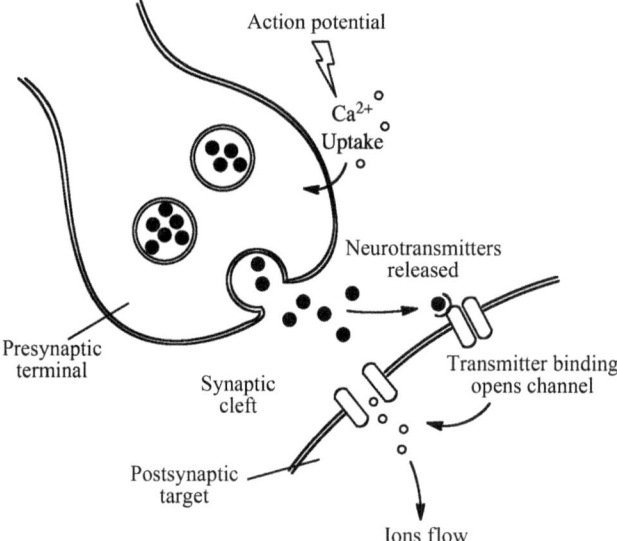

Figure 2.5 Synaptic junction between beurons.

Figure 2.4. There are three parts—soma or cell body, dendrites and axon. The soma is the central component of the cell which contains the nucleus where protein synthesis occurs. The axon is a projection that extends some distance in comparison with the soma, depending on the particular function of the cell. Close in proximity to soma is the axon hillock, where there is a high density of Na^+ channels. The axon itself contains structures called nodes of Ranvier (constrictions) and is 'coated' with a myelin sheath. Filament-like dendrites in great numbers, and displaying a high level of branching, emanate from the soma. Signals from other neurons are received by the soma *via* the dendrites and signals from the axon terminal are transmitted to other cells. The signaling mechanism has components of electrical and chemical processes which occur *via* the synapse (Figure 2.5). (The human adult brain possesses some 10^{14} synapses!) In this structure, close proximity is exhibited by bilayer lipid membranes which, intrinsically, display high electrical resistance due to the close-packing of the lipid molecules. The membranes contain protein assemblies which constitute ion channels and ion pumping configurations. The channels can be gated by voltage effects or by chemical stimulation *via* molecules termed neurotransmitters. The electrochemical nature of the signaling process is described below.

With respect to type, neurons, which are highly variable in nature (see Figure 2.6 for examples) can be categorized *via* number of different classifications. However, the most important parameters are location in the nervous system, morphology and function. At the first level, neurons exhibit unipolar, bipolar or multipolar character. This is not an electrical term, but is employed to describe the positioning of dendrites with respect to the soma and axon. Additionally, in general terms of behavior, cells that relay information about their environment, such as olfactory cells, for transmission to the CNS are called afferent neurons. Those that orchestrate signals originating from CNS are termed efferent neurons. There are also interconnecting neurons between these two types of structure. At the molecular level neurotransmitters, through interaction with particular receptors, can excite, modulate or inhibit activity in target cells. This leads to a neural classification based on the chemistry of generation of neurotransmitters—the terms end in '-ergic' as a descriptor. Examples are cholinergic (acetylcholine production), dopaminergic (dopamine) and serotonergic (serotonin) neurons. The biochemical activity of these types of neuron and other neurotransmitter-producing cells is summarized in Table 2.3.

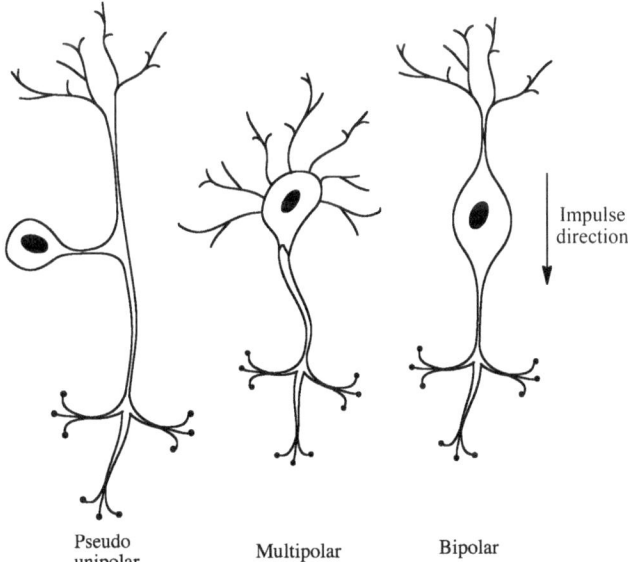

Figure 2.6 Neurons with different polarity.

Table 2.3 Examples of neurotransmitters and their actions.

Neurotransmitter	Action	Structure
Acetylcholine	Excitatory, triggers muscle contraction Stimulates hormone secretion in CNS	
Dopamine	Brain—regulation of motor behavior Connected to motivation and emotion	
GABA	Fast inhibitory effect at synapses in the brain	
Glutamate	Fast excitatory effect at brain and CNS synapses	
Norepinephrine	Import mat with respect to attentiveness, sleep, emotion Released as a hormone into the blood steam	
Serotonin	Regulates appetite, sleep, mood, behavior and muscle contraction	

GABA, γ-aminobutyric acid.

2.4.2 Action Potential and Electrical Conduction

An electrical potential difference exists across the membranes of most biological cells including neurons.[18] By convention the value of the potential, which varies between about −10 and −100 mV, is given a negative sign to recognize that the interior of the cell is charged negatively with respect to the exterior. This parameter is most often referred to as the 'resting' or 'steady state membrane potential'. The cell membrane is permeable to Na^+, K^+ and Cl^- *via* specific channels rather than simple pores. The chloride concentration in the ECM is relatively high, resulting in diffusion into the cell, but the negatively charged interior opposes this process by electrical repulsion with the end result of an equilibrium situation. The reverse situation exists for the potassium cation distribution. With respect to Na^+ both the electrical and concentration gradients are in the same direction—inward. These are passive considerations, but in reality active transport of both sodium and potassium cations across the membrane is involved, which tends to keep the interior concentration of the

The Cell-Substrate Surface Interaction

ions constant. The mechanism responsible for the transport of sodium out of the cell and potassium in the reverse direction is termed the 'sodium–potassium pump'. The pump is fuelled by enzymatic hydrolysis of ATP.

The impulse stimulus of the axon of a neuron results in a number of electrical potential changes collectively referred to as the 'action potential'. After a latent period a depolarization of about 15 mV occurs; this is followed by a further rapid depolarization. The point at which the latter occurs is often tensed the 'firing level'. The potential then rises above zero in an overshoot effect before eventually returning to the resting potential. The time sequence of the events associated with the action potential is summarized in Figure 2.7. The signal represents an 'all or none' process with respect to firing.

We can now summarize the effect of a stimulus on the overall behavior of the (unmyelinated) axon. Polarization of the membrane involves an excess of positive charge on the outside and prevalence of negative charge on the inside of the membrane. This polarization is eliminated during an action potential, with the result of increase of flow of positive charge into the negative sink, which in turn depolarizes the membrane ahead of the action potential. This process represents a self-propagated response along the axon and circular flow of current.

Myelinated axons with the insulating sheath and nodes of Ranvier in place exhibit a similar circular (saltatory) flow of current to that described above. However, in this case, the depolarization event jumps for one node to the next, resulting in a propagation of the action potential which is many times faster than for uninsulated axons. Although in principle axons can conduct current in both directions, in living animals they are unidirectional only to their

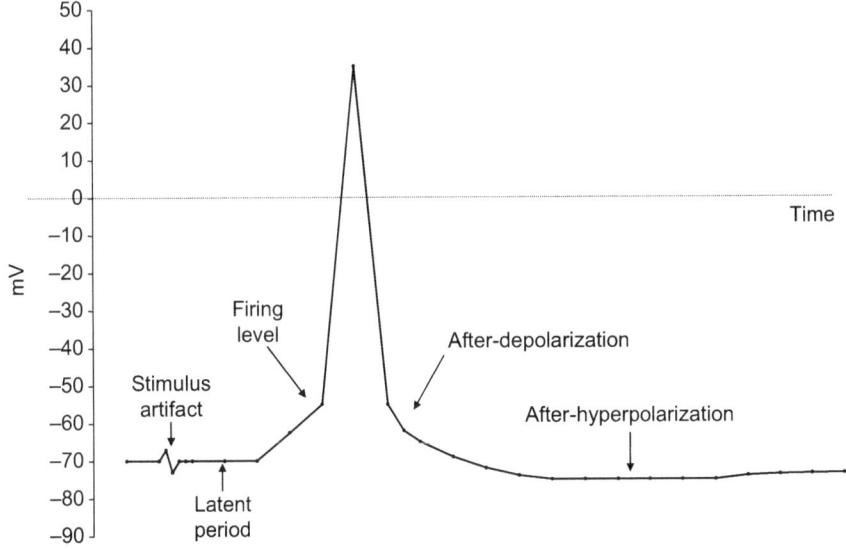

Figure 2.7 Idealized time course of cellular action potential.

termination at the synaptic junction or receptor. As noted above the hydrolysis of ATP yields the energy required for operation of the pump, but the rate of metabolism increases significantly during the propagation of the action potential. Obviously, in nerve tissue of varying types the overall behavior will reflect the compounding of a vast number of action potentials and signals, the most complex of this being the human brain.

2.5 Cell Adhesion, Growth, Guidance and Proliferation on Substrates

2.5.1 General Considerations

The cell–substrate interaction has been the subject of intense research for many years, with the most activity being directed at the biomaterials sector. The main focus of this research has been on producing a 'scaffold', whether of synthetic or natural material, that allows the structural support and maintenance of healthy cells when in contact with such a material. Research with the aim of 'repair' in mind is often associated with the marketing term 'regenerative medicine', although the developments can hardly be described as regenerative in the strictest sense of the word! In the biomaterials world, the biodegradability of the substrate is, additionally, an important factor. The reason for this lies in the potential inadvisability of the presence of foreign entities in the body. As outlined at the end of this chapter, there are medical consequences as a result of the deleterious effects of surfaces on blood components, including various cells.

There are two strategies to achieve cellular adhesion and growth—substrate chemical modification and sub-micrometer patterning of the surface topography. Wells, pillars and grooves of different dimensions are fabricated using standard microfabrication techniques, including photolithography, wet etching, reactive ion etching and laser fabrication. Chemical modification means changing the surface composition and properties by chemically binding different proteins and molecules that promote cell adhesion. An unexpected finding is that cellular response to engineered surfaces is specific to the cell type. This result clearly implies that there is no general method that fits all cells types, including neurons.

Neuronal cells are functionally and morphologically unique among cells for various reasons. They are polarized, possessing receptive dendrites on one end and axons with synaptic terminals on the other. Impressively, the brain is not a static structure but it is capable of neuroplasticity, an essential process for cognition and learning. It is possible to induce physiological changes that lead to anatomical changes, including pruning of preexisting connections and growth of new connections. These amazing properties of the brain are explored in the research on neuronal prosthesis for neuronal recovery. Neurite guidance, especially axonal guidance, is one of the key processes during embryogenesis and is explored in neural tissue engineering. Neural cells find their counterparts for signal delivery with the aid of an expanded terminal structure called the growth cone at the end of the axon shaft. The growth cone has some specific features consisting of filopodia and lamellipodia developed by F-actins and microtubules. Those cellular cytoskeletons change their spatial organization inside the growth cone continuously *via* external cues that mostly originate from the substratum over which cells adhere by a phenomenon known as contact guidance. Neuroscientists have also observed that glial cells in the central nervous system provide natural physical cues to migrating neurons. Basement membranes, which are a natural substrate for cells in different parts of the body such as the corneal epithelia, possess a complex, three-dimensional (3D) topography consisting of nanometer-sized features. The topographic cues of the ECM play vital roles in cellular behavior, adhesion, spreading, migration, proliferation and differentiation. Researchers have tried to mimic or amplify the topographical effects of the cell's native environment by creating artificial substrata designed with microstructures and nanostructures that stimulate neurons to grow around such structures.

As mentioned above, the goal of making chemical and topographical modifications on a substrate material is to enhance biocompatibility as well as to modulate cell adhesion. In this respect it is crucial to recognize that an implantable electronic device, for example, will be comprised of different components made of various materials—electrodes, insulators, encapsulation materials and so on. Material use will be highly variable ranging across polymers, glass, indium tin oxide (ITO), metals, silicon, quartz and sapphire.

The main purpose of the rest of this chapter is to review concisely, with an emphasis on neurons but not restricted to this cell, the various approaches that have been employed to attempt the retention of biological integrity, healthy growth and proliferation. The spatial guidance and patterning of neurons on substrates as it pertains to the formation of networks is also discussed in Chapter 4.

2.5.2 Bare Substrates

The most straightforward approach has been to simply attempt the introduction and growth of cells on either bare inorganic, surfaces (*e.g.* plain metals, silicon) or organic materials such as polymers. Often the aim is a comparison of bare surfaces with treated substrates in order to investigate possible methods for enhancing biocompatibility and hemocompatibility (*e.g.* reduction of thrombogenicity). We outline some examples here (morphology, physical and chemical patterning of these substrates is discussed later).

One of the materials heavily employed in devices associated with bioanalysis, in particular microfluidic and so called 'lab-on-a-chip' structures, is poly(dimethylsiloxane) (PDMS)[19–21] and polymers belonging to a similar chemical family. This material is thought to possess advantageous properties in terms of biocompatibility.[22] The polymer is also low in terms of permeability to water and displays low electrical conductivity.[23] For these reasons it has been very much the material of choice for those working with microfluidic channel-based devices.

In a comprehensive study, Whitesides and co-workers[24] examined the behavior of a number of mammalian cells on various forms of the polymer. Although polymer samples were studied with respect to composition, experimentally, surface adsorbed fibronectin was employed to aid cell 'adhesion'. Accordingly, there is the caveat attached to this work in that the results are not necessarily governed entirely by chemistry conducted at a 'bare' surface. Four different types of cells were investigated—primary human umbilical artery endothelial cells (HUAECs), transformed 3T3 fibroblasts (3T3s), transformed osteoblast-like MC3T3-E1 cells and HeLa (transformed epithelial) cells. Several polymer slab substrates were prepared for exposure to the cells with an emphasis on the ratio of base-to-curing agent employed. Polymer samples where the surface was treated with extracting solvents and subjection to plasma treatment were also examined. The growth of cells was detected *via* fixing and fluorescence spectroscopy, following surface characterization. In overall terms an attempt was made to correlate the growth of cells on the various substrates with surface chemistry, as influenced by polymer fabrication and stiffness *via* Young's modulus assessment. There was some dependence on cell type but stiffness appeared to have a marginal effect. The role of the supposed underlying layer of fibronectin in this study remains unclear, at least in terms of the surface chemistry component. In other words, does the behavior of cells on the different polymer surfaces simply reflect the nature of the specific fibronectin interaction?

With regard to metals there has been particular interest in medical-grade stainless steel surfaces and how this material interacts with cells. The research has been largely spawned by the introduction of stent technology for the treatment of coronary disease.[25] (In the last section of this chapter we revisit this area with regard to the medical consequences of implantation of foreign substrates.) There are numerous examples of this sort of work involving the use of electrochemical pretreatment of stainless steel in order to enhance biocompatibility. These are reviewed in ref. 26. In one example, stainless steel (316LS) was subjected to passivation by an electrochemical process with the goal of examining responses of fibrinogen, platelets, endothelial and smooth muscle cells.[27] Interestingly the electrochemical treatment reduces the accumulation of platelets on the steel surface (by around 30–50% depending on exposure time). Also there was evidence of less activation of platelets by the modified surface compared with the bare substrate.

The effect of cells on steel, as distinct from what the metal surface does to cells, has also been studied. An example is the investigation of the corrosion of steel by osteoblasts.[28] (These cells originate from marrow-derived monocyte precursors and are multinuclear in character.) The study found that the cells could be cultured successfully on bare surgical grade stainless steel. After several days of exposure to the cells, corrosion-instigated pits were observed on the steel wafers by scanning electron microscopy (SEM). Analysis of the culture supernatant liquid revealed the presence of expected metal ions derived from the metal (*e.g.* Ni^{2+} and Mn^{2+}) which were not present in control samples. Although the precise mechanism of the corrosion remains obscure the authors specified that it must have an 'electrochemical' basis. From the medical point of view it is interesting that the steel–cell system resulted in the generation of cytokines, which are pre-inflammatory species.

Another recent study dealt with an effort to prevent the adhesion and proliferation of cells on medical grade steel by coating the metal with polymer films.[29] Again the culturing of murine fibroblasts and human umbilical vein endothelial cells on the metal was successful and polymer fibrils were capable of reducing the numbers of such cells. The authors pointed out such coatings could have advantages in terms of anti-thrombotic behavior compared with other methods for biologically passivating steel, (see the last section of this chapter).

A final example that pertains to bare surfaces is some of our own research on the behavior of various cells on gold. The particular interest in this case is the label-free detection of the behavior at a surface of aortic smooth muscle cells (ASMC)[30] and neurons[31] by acoustic wave physics which involves the use of a gold electrode. (The physics and its relevance to neuron behavior in particular is the subject of a later discussion in the text.) The ASMCs were harvested from embryonic rat aorta and cultured by standard methods. The cells were introduced to the on-line detection system *via* a flow-injection configuration. The response of the acoustic sensor showed clearly that AMSCs attach to bare gold successfully and are not removed by washing (Figure 2.8). Quite surprisingly the bare metal surface yielded the best conditions with respect to both surface capacity and fastest kinetics of attachment compared with laminin and fibronectin coated electrodes (on-line). This observation allowed the study of cell detachment by trypsin and *in situ* destruction by various reagents.

For experiments with neurons, hypothalamii were harvested and dissected from mouse embryo and grown in primary culture. The cultures were infected with a replication-deficient retrovirus. The minced hypothalamic cultures were then subcloned to yield useable clones for experimental work. Immortalized neurons were cultured directly onto the gold electrode surface to a confluence of 80–100% (Figure 2.9). Using this configuration the behavior of the cells was examined in terms of various parameters and effects. These included cell adhesion and proliferation, depolarization and the response to stimulants such

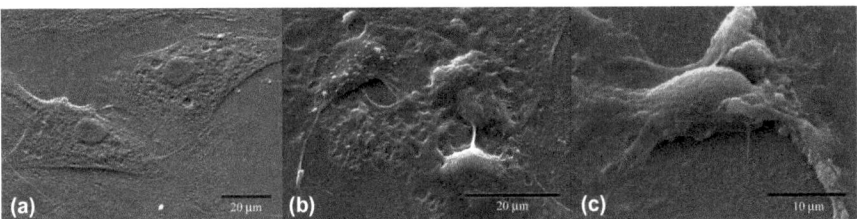

Figure 2.8 Images of thickness shear mode (TSM)-surface attached cells under SEM. (a) Structure of cells before interacting with hydrogen peroxide; (b) and (c) Morphology of cells after the addition of hydrogen peroxide.

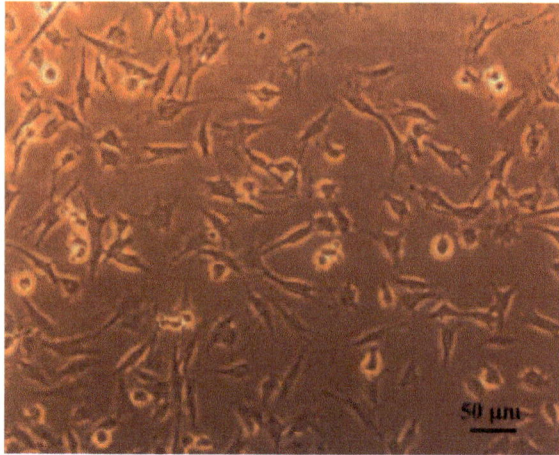

Figure 2.9 Optical micrograph of mouse N-38 hypothalamic neurons cultured on TSM electrode.

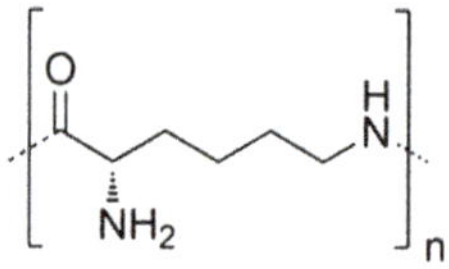

Figure 2.10 Basic repeat unit of PLL.

as glucagon and nanoparticles. As was the case for ASMCs, results for the bare surface were compared with coated devices. The results of this work centered on neurons are discussed in more detail later.

2.5.3 Polypeptide Coating

As implied in the previous section, a myriad of different coatings and films imposed on surfaces have been employed to attempt enhanced adhesion of cells to a substrate with varying degrees of success. By far the most prevalent molecule that has a long history of such use is poly-L-lysine (PLL), a homo-polypeptide of some 30 residues (L-lysine). Not only has the molecule been applied to the area of cell surface attachment, but it also appears to figure prominently in DNA and protein microarray technology for the glass surface binding of protein and nucleic acid probes for multiplexed assay of biochemical interaction.[32,33] The use of PLL is so ubiquitous that a number of commercial preparations are available.

The molecule is positively charged (Figure 2.10) and is considered to bind to surfaces *via* electrostatic chemisorptive forces. Over 30 years ago it was suggested that various cells could be attached to a PLL-coated substrate *via* the polyanionic functional groups present at the cell surface. The specific application at that time was connected to sample preparation prior to study by electron microscopy.[34] The following years have seen literally hundreds of publications where PLL is employed for cell attachment to surfaces of ceramics, hydrogels, plastics, metals, *etc.* An excellent example in this area was the study of human K562 erythroleukemic cells grown in PLL-coated culture containers.[35] Interestingly, the authors of this work established that changes in both membrane conductivity and permittivity were apparently altered by the cell–PLL interaction. Redistribution of cellular species after exposure to PLL was investigated using flow cytometry and immunofluorescence microscopy. The results seem to indicate that such redistribution may, in part, be responsible for cellular adaptation to the new growth environment of K562 cells and for the variations in membrane electrical properties observed.

More recent rimes have seen attempts to adhere cells to substrates in a three-dimensional fashion. In this work, thermoformable polymer films were fabricated to produce micro-structured scaffolds, *e.g.* curved and μ-patterned substrates, rather than the more conventional planar system. The surface of a poly-lactic acid membrane was coated with a photopatterned layer of PLL and hyaluronic acid (VAHyal) to attempt spatial control over cell adhesion. Human hepatoma cells (HepG2) and mouse fibroblasts (L929) were used to demonstrate so-called guided cell adhesion.[36]

Closer to home with regard to this text, PLL has also been used recently in conjunction with polyethylene glycol (PEG) based hydrogel to fabricate a superior neuro-electrode interface.[37] The thrust of this work was the fabrication of neural prostheses, which is an extremely active area of research connected to the treatment of lost neural function (discussed in more detail later in the text). In order to mimic natural neural tissue it is important that the prosthetic device exhibits appropriate properties, *i.e.* biocompatibility with respect to mechanical and surface chemical behavior; as alluded to previously, hydrogels possess ideal characteristics from both perspectives. In particular, the use of polyethylene glycol is extremely popular in view of its well-established biocompatibility properties. The polymer is known to be resistant to cell and biochemical macromolecule non-specific adsorption, conferring reduced biological damage at the biospecies–substrate interface. (The reason for this effect remains obscure, although the role of trapped water has very often been ascribed to contribute to the mechanism.) Accordingly, a technique was reported which described enhancement of the interface between polymeric brain mimetic coatings and neural tissue using PLL. Polymer-modified PEG-based hydrogels were synthesized, characterized and shown to promote neural adhesion using a PC12 cell line. In addition, it was observed that the polymeric materials adhered to electrodes for at least four weeks. These results suggest that PLL–PEG hydrogel biomaterials are biocompatible and can enhance stability of chronic neural interfaces.[37]

2.5.4 Extracellular Matrix Proteins and Derived Peptides

Given the role played by the ECM proteins and other macromolecules with respect to cellular adherence, progenitor development, communication and differentiation,[7] it is not surprising that an enormous amount of research has centered on the modification of inorganic and polymer substrates with these molecules prior to experiments with cells. Most work has focused on laminin, collagen and fibronectin where either the whole protein complex (or cocktails thereof) or peptide sequences are 'grafted' onto a substrate. Briefly, the laminins are a family of glycoproteins that are responsible for structural scaffolding in tissue. They are trimeric molecules comprised of α-, β- and γ-chains present in five, four and three genetic variants. Fifteen laminin structures have been found where variability is defined by the chain composition. The proteins connect to other ECM macromolecules and membrane components to form cross-linked scaffolding. Collagen is massively present in connective tissue in mammals. It is composed of three polypeptide chains with the molecule being incorporated into a larger fibril structure. The protein is characterized by possessing high concentrations of glycine, proline and hydroxyproline, which confer properties with respect to connectivity.

As for the case of polypeptides, the 'grafting' or attachment of ECM protein or peptides to particular substrates is based conventionally on either non-covalent or stronger covalent interactions. Hydrophobic force, van der Waals forces, hydrogen bonding or electrostatic forces are the basis of protein adsorption by non-covalent interactions. Modification by such binding of coating materials on the surface is advantageous because of their ease of application since no chemical modification is required prior to immobilization. The drawbacks are protein denaturation due to uncontrolled interactions between the proteins and the surface, and undesired protein desorption during the study period. A more stable means of protein immobilization is to link a protein to the surface covalently *via* a chemical bond.

Widely used peptides derived from ECM proteins that are used for neuron attachment are GDPGYIGSR, GQAASIKVAV, GRGDS and RHSRN. An excellent example of this type of chemistry is the use, similar to the hydrogels mentioned above, of a three-dimensional polymeric matrix specially synthesized from a methacrylated dextran and aminoethyl methacrylate in the presence of PEG.[38] This leads to a macroporous structure displaying high equilibrium water content. The presence of the amino functionality was assessed semi-quantitatively *via* a simple ninhydrin color test. Covalent binding of peptides to the $-NH_2$ group present on the copolymer was achieved *via* a linker. The latter activated the surface group in a somewhat analogous manner to that employed in biosensor technology discussed in Chapter 1. Peptides studied were CRGDS or a mixture of CDPGYIGSR and CQAASIKVAV. Significantly enhanced growth of primary embryonic chick dorsal root ganglia on peptide-treated polymers was observed (Figure 2.11). Despite reaching the conclusion that ECM-derived peptides on polymers could provide an ideal scaffold for neural regeneration, the authors appeared disappointed by the level of neural attachment and performance obtained in their experiments.[38]

Guidance of neurite growth using ECM-based peptides incorporated again in a three-dimensional polymer hydrogel has been attempted.[39] In this work the polymer hyduronan was reacted with *S*-2-nitrobenzyl cysteine using the

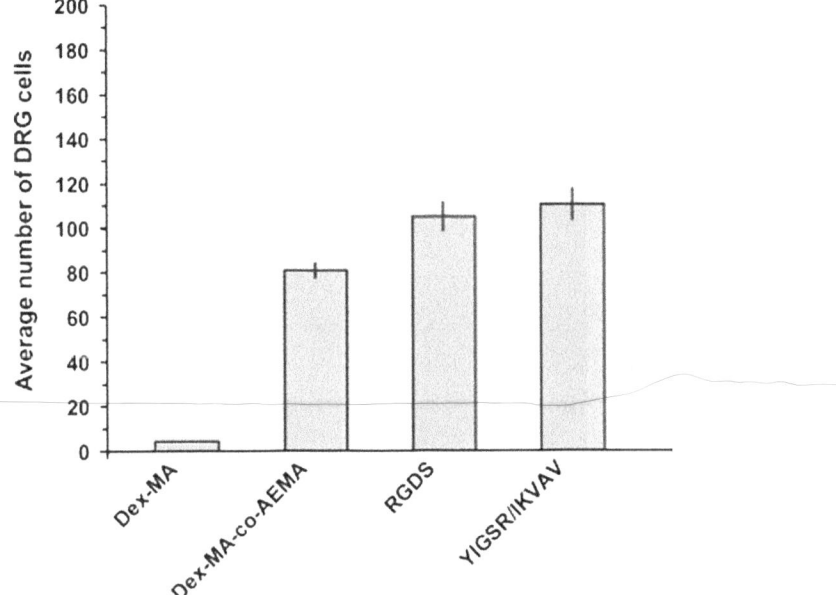

Figure 2.11 Average number of dorsal rat ganglia cells adhered on various peptide-modified surfaces using a base hydrogel. The two columns to the left represent values for pure polymers whereas those to the right are for peptide modified hydrogels.
(Reprinted by kind permission of Elsevier, BV.)

standard surface ethyl(dimethylaminopropyl) carbodiimide–N-hydroxysuccinimide (EDC-NHS) activation chemistry. (The chemistry behind the latter is outlined in Chapter 1.) NMR spectroscopy was employed to ascertain the composition of the resulting hydrogel. Following exposure to ultraviolet (UV) light in order to create cylindrical channels, GRGDS oligopeptides could be attached and concentrated within the channels; the latter are shown in Figure 2.12. The next step was to allow the UV/peptide-modified hydrogel to interact with neural cells derived from rat dorsal root ganglia, the idea being to examine the possibility for peptide-guided neurite growth following cell migration within the channel volumes. The remarkable result was obtained that the GRGDS gradient across the x–y plane of the biochemical channels affected neurite outgrowth such that neurites extended preferentially up the concentration gradient towards the central core of the peptide channel. The authors pointed out that the result constituted the first time that guided neurite in such a fashion had been achieved.

A similar approach was used by Hynd and co-workers[40] with respect, also, to directed cell growth on a hydrogel. In this case the hydrogel was based on photo-polymerized acrylamide. The attachment of biotin-tagged ECM proteins to the hydrogel was achieved by co-polymerization of the polymer with streptavidin. Micro-contact printing was used to pattern the proteins such as laminin on the hydrogel surface. It was observed that both LRM55 astroglioma

Figure 2.12 A representative longitudinal image of green, fluorescently labeled oligopeptide channels constructed from a series of XY cross-section micrographs taken at 20 μm intervals over a depth of 0.5 mm (contrast adjusted to clearly view the isolated regions).
(Reprinted by kind permission of the Institute of Physics Publishing.)

and primary rat hippocampal neurons adhered to areas patterned with biotin-conjugated proteins. Fluorescence and bright-field modes of microscopy were used to measure cell morphology on modeled surfaces. LRM55 cells were found to attach to protein-stamped regions of the hydrogel only.

There have been a number of other studies of ECM proteins attached to polymers of various types, but particularly hydrogels, and these are reviewed in ref. 41.

A more orchestrated approach to look at cell attachment to ECM proteins and peptides has been the use of spatially modified substrates such as glass and silicon (the field being beautifully reviewed in ref. 42). The idea that the surface spacing of ECM-based RGD peptides is important in terms of cell adhesion dates back to the 1990s.[43] In this early work the synthetic peptide Gly-Arg-Gly-Asp-Tyr (GRGDY) was attached to a glass surface with different concentration of peptide being employed to purportedly 'space out' the molecules on the surface. Using fibroblasts it was shown spreading of cells occurred at a minimum spacing of 440 nm, whereas focal contact formation occurred at 140 nm. Although these results on a two-dimensional (2D) substrate indicated the importance of peptide spacing, it should be emphasized that little or no surface characterization was included in the experiments.[43]

Much later, a more rigorous experiment was published in which nano-patterning of RGD peptides was employed to study cell spreading.[44] Copolymers in micellar structural form contained gold nanoparticles which were then deposited on glass slides. The polymer was removed for this system via plasma treatment, leaving gold nanoparticles of 5–8 nm diameter. The gold particles were organized in hexagonal patterns and are separated by 28–110 nm. The space between gold nanoparticles was then covered with PEG (molecular weight 2000), supposedly to prevent cell and protein adhesion. The interaction between Au and sulfur was used to attach the thiolated cyclic peptide c-RGDFK to the gold particle. Cellular adhesion experiments were conducted using integrin-transfected fibroblasts. The results of this work indicated strongly that, when exposed to the nanoparticle modified surfaces, the cells were able to 'recognize' spacing effects. Using a phase-contrast microscopy protocol it was shown that fibroblasts spread well for the case where nanodots were spaced by 50 nm, in contrast to the situation with 110 nm spacing where limited spreading was observed. In addition to this observation it was noted from the microscopy that quiescent cells are rounded whereas migrating cells exhibit a polarized shape. The role of integrin $\alpha_v\beta_3$ during adhesion to RGD nanopatterns at different spacing was also examined via a technique which involved the antibody-based blocking of receptors prior to seeding cells onto the surfaces. The number of attached cells was evaluated one hour after seeding using formaldehyde fixation and toluidine blue staining. Interestingly, the number of cells adhered to both spacings were almost identical after such a blocking procedure. In addition to these experiments the authors also studied the recruitment of integrin-associated molecules in terms of their dependence on the distance between integrin ligands. Example images from this type of research are shown in Figure 2.13.

It is not only the fundamental spacing between RGD sites that can influence cellular behavior. Experiments have indicated that the actual ordering of nanopatterns on surfaces can mitigate cell adhesion. In an analogous approach to that described immediately above, ordered and disordered nanopatterns using the gold particles were produced on a supposedly bio-inactive substrate.[45] The various types of ordering were instigated through the use of an 'interference reagent' (a copolymer) during nanopattern formation and cellular adhesion studies involving osteoblasts. Ordering was characterized through an order parameter in addition to particle diameter and spacing on the surface, as evidenced from atomic force microscopy. An important aspect of this study was that the nature of the nanoparticles meant that only one integrin complex could be attached to an RGD-treated particle. After allowing the fluorescent-labeled cells to interact with the various substrates, the number of attached cells and area occupied by osteoblasts were determined. It was found that the number of cells and area occupied on ordered patterns decreased with increasing distance between nanoparticles. This was not the case for disordered patterns, the result being ascribed to the greater variety of ligand (peptide) densities. Moreover, the crucial nature of interaction or clustering of integrin moieties in the cell membrane is more likely to be exhibited in the case of disordered surface nanodots. This effect is shown schematically in Figure 2.14.

Although gold–sulfur functional group SAM chemistry outlined in Chapter 1 was proposed some years ago as a useful structure for studying RGD

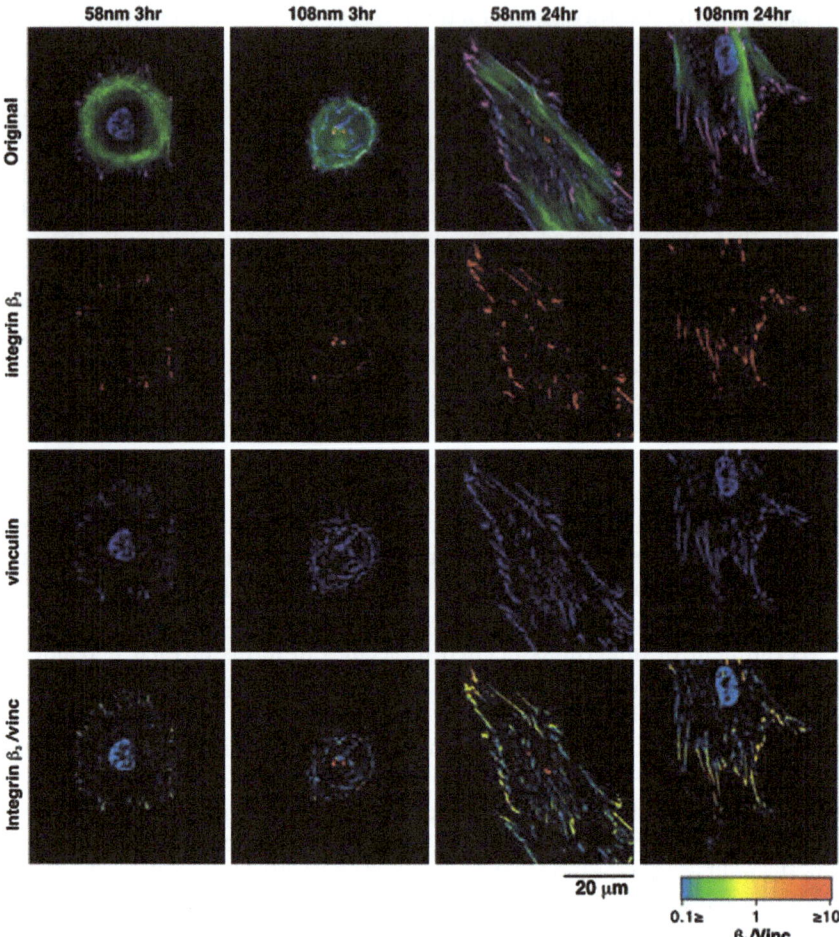

Figure 2.13 Immunostaining and fluorescence ratio images (FRI) of focal adhesion proteins. REF52 cells transfected with GFP-integrin β3 were fixed and immunostained with primary antibody against vinculin, followed by Cy5-conjugated secondary antibodies. Actin filaments were visualized with phalloidin-TRITC. Cells on 58 nm and 108 nm RGD nanopatterns were observed at 3 h and 24 h after plating. The rows present the images with integrin β3 in red, vinculin in blue, and actin in green. The last row shows the ratio between integrin β3 and vinculin intensities. FRI are presented in a spectrum scale as indicated in the lookup table.
(Reprinted by kind permission of Cell Press, Biophysical Society.)

spatiality,[46] the lack of robustness of such films has been cited as constituting a serious problem for this approach.[47] An attractive alternative is provided by hydrosilylation chemistry, which results in the attachment of an organized monolayer with strong Si–C covalent bonds.[48] An example of RGD attachment to such a surface is the production of a mixed monolayer on silicon of oligo (ethylene oxide) species.[49] In this experiment the main focus was the

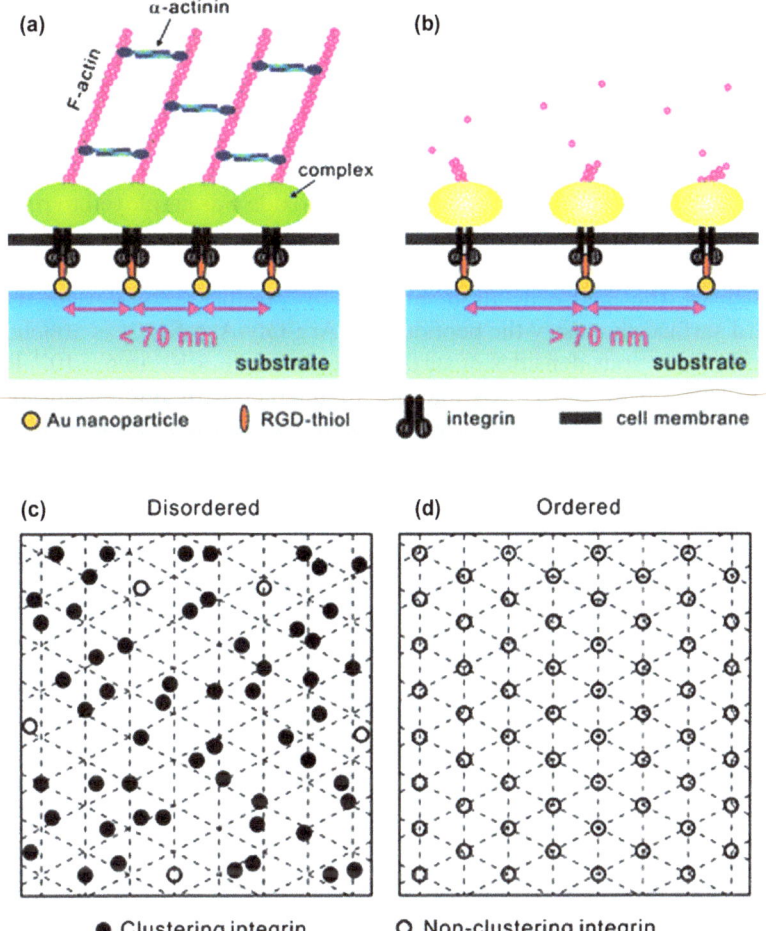

Figure 2.14 Sketch of integrin clustering and subsequent Focal adhesion (FA) formation regulated by different c(-RGDfK-)-thiol ligand nanopatterns. The spatial arrangement of the original Au nanopatterns well reflects the c(-RGDfK-)-thiol ligand lattice and, thus, the final lateral positioning of single integrins during cell adhesion. (a) A spacing of <70 nm between two neighboring c(-RGDfK-)-thiol ligands results in effective integrin clustering and FA complex formation, followed by the formation of the F-actin cytoskeletal network (only some of the intracellular molecules were depicted here). (b) In contrast, a spacing of >70 nm, as such, results in neither integrin clustering nor FA complex formation. (c, d) It was presumed that all integrins that potentially bind c(-RGDfK-)-thiol ligands over each nanopattern could be classified as clustering integrins (black disks); non-clustering integrins (white disks) resulted from inter-distances above the critical value. Even at a global average interligand spacing of >70 nm, a disordered nanopattern still displayed some clustering integrins, which was not the case for ordered patterns with interligand spacings of >70 nm.
(Reprinted by kind permission from the American Chemical Society).

study of biocompatibility by randomly distributed but spaced RGD moieties. There was no attempt to orchestrate ordered patterns with respect to peptide spacing as mentioned above.

In contrast, Gooding and colleagues[50] have specifically pinpointed the fact that etching of silicon, normally used to produce electronic devices, can be utilized to provide combined topography and RGD presentation to cells. Wet chemical etching of silicon with potassium hydroxide was employed to produce random pyramidal structures on both the nano- and micro-scale of the semiconductor surface. Following removal of the oxide layer the silicon wafers were subjected to the hydrosilylation process to produce monolayers. Using standard surface chemistry the peptide, Gly-Arg-Gly-Asp-Ser, was attached to the surface in various densities of surface population (6×10^2 to 6×10^{11} RGD mm^{-2}). This surface was then allowed to interact with fluorescent-labeled bovine endothelial cells. Interestingly it was found that the flat or nanoscaled surface adhered more cells than was the case for rougher surfaces. Cell spreading was controlled more by the population density of the peptide than by substrate morphology. The important conclusion was reached that initial contact of cells with a substrate may be controlled by the topography whereas the engagement of cell surface receptors is dominated by the surface chemistry, *i.e.* RGD surface density.

In the context of the results outlined above, Gooding and colleagues[51] had previously commented on how cells migrate on the micro- and nanoscale with particular reverence to oncology. In this earlier paper they reviewed the stages of cell migration subsequent to receiving ECM cues—polarization, protrusion, traction and disassembly. These processes can be instigated by various chemical signals, such as hormones and growth factors, where the spatial chemistry of ligand binding to cellular receptors is critical. The possibility of fabricating microfluidic systems (see Chapter 4) and of incorporating nanostructures on surfaces offers great potential for examining how cells begin the migration process. This is important in biological processes such as metastasis.

Given the success of patterning both by nano-gold particles and silicon nanofabrication protocols, it appears that the recent silanization adlayer chemistry introduced by Thompson *et al.*[52] and described briefly in Chapter 1, offers an attractive alternative to hydrosilylation. In this case covalent bonds to –OH groups on a variety of surfaces *via* simple trichlorosilane reactions could clearly be applicable to SiO_2 and, therefore, patterning. The material advantage over other systems is that the adlayer yields highly biocompatible substrates in addition to the basic requirement for the possibility to attach orchestrated patterns of peptide and such ligands.[53] The technology was developed with biosensor devices in mind, but the avoidance of surface protein adsorption also offers possibilities with respect to biocompatibility in biological fluids.

2.5.5 Substrate Morphology

In the previous section we referred to research on the influence of surface-attached RGD and patterned peptide spacing on cell–surface behavior.

The Cell-Substrate Surface Interaction

There has also been very significant interest in the general role played by physical micro- and nanoscale surface structuring of materials (of varying degrees of order) with respect to cellular response. This property is especially critical with respect to materials employed in medical implant and extracorporeal device technology. The literature is replete with studies of a wide variety of cells. We concisely review this work here, but leave consideration of such research on neurons, including surface patterning, until later chapters.

An extensive battery of techniques has been employed to create 2D and 3D nanostructured features such as photolithography, focused ion beam lithography, e-beam lithography, nano-imprint lithography, interference lithography, reactive ion etching, glancing angle deposition, physical vapor deposition, electro-spinning, self-assembly patterning, colloidal lithography, polymer de-mixing, co-block phase separation, two-photon polymerization and chemical etching or oxidation. Examples of these approaches are given in references 54 and 55. Features instigated in substrates include gratings, posts, pits, and island geometries in both the micrometer and nanometer range (see, for example, Figure 2.15). However, these techniques each have their own merits and limitations which can vary with respect to material, geometry, cost and surface area coverage. If the main application is biomaterials research, important considerations will be processing over relatively large surface areas and overall cost-effectiveness. These criteria stand in sharp contrast to the fundamental research on surface–cell behavior alluded to in the previous section.

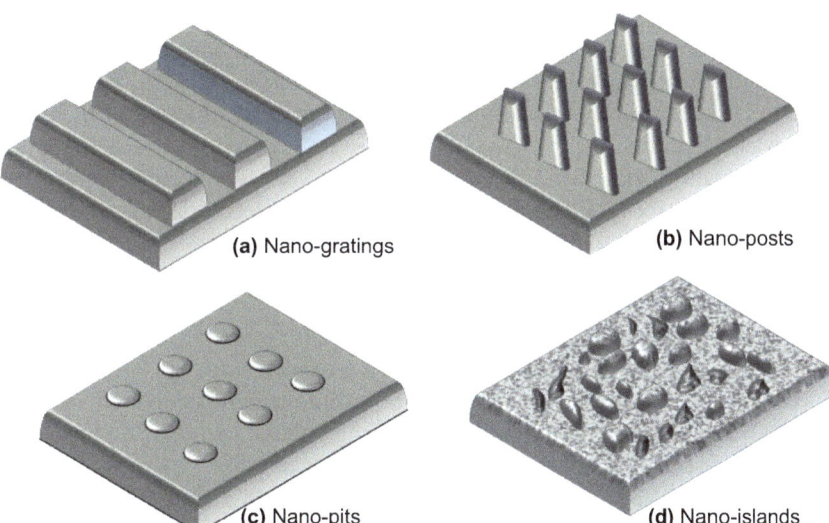

Figure 2.15 Schematic of typical surface topographies produced for study of cell adherence and behavior.

2.5.5.1 Morphology and Response of Endothelial Cells

Previous studies have reported that endothelial cells (ECs) can respond viably to surface topographical features in the micrometer, submicrometer and nanometer ranges. For example, Palmaz et al.[56] found that on NiTi surfaces discontinuities at the microscopic scale influenced the conformation and motional dynamics of migrating ECs. Migration speeds increased (up to 64%) on surfaces with gratings ranging from 3 to 22 μm compared with flat control surfaces. Patterning of Ti surfaces involving periodic arrays of gratings, with widths and spacings ranging from 750 nm to 100 μm using a plasma-based dry etch technique, have been used to study EC adhesion, proliferation and morphology.[57] In this work, it was observed that ECs on nanoscale patterned Ti surfaces were oriented and displayed enhanced cell function compared with smooth Ti surfaces and random nanostructured Ti surfaces. A number of articles discussing a variety of substrates have reported that the line grating geometry results in enhanced adhesion and orientation of ECs.[58–60] A frequent comment emanating for this sort of study suggests that the line grating is analogous to native endothelium. However, there have also been a number of reports which indicate that such patterns result in reduced cellular proliferation. The reason for this observation is unclear at the present time and clearly requires further investigation.

The effect of nano-island geometry, especially for experiments involving polymer surfaces, on EC behavior has also been extensively investigated. Interestingly, both increased and reduced spreading of ECs on such patterns have been reported.[61,62] In overall terms these studies highlight the fact that quite slight changes in feature height produce large changes in EC behavior. In addition, surface chemistry clearly has an effect on EC response because there is significant variation in cellular response associated with the various polymer mixes employed in the research.

A limited number of studies on the response of ECs to nano-posts have been reported. For example, Kim et al.[63] demonstrated that nano-post PEG surfaces fabricated using capillary lithography enhanced the focal adhesion of ECs. This result was attributed simply to an increase in material surface area and adhesion sites for cells. Recently, Zawislak et al.[64] showed the development of ECs on 3D nano-post silicon surfaces with a depth of 10 μm, a periodicity of 6 μm and a diameter of 0.15 μm. In a sub-confluent layer, the cells impaled themselves on the pillar to the extent that even some of the cell nuclei were penetrated by the pillar tips.

Given the great importance of new biomaterials in the world of stent technology it is not surprising that there has been a concentrated effort to study the behavior of ECs on substrates that display random nanostructured features. For example, increased EC function (including collagen and elastin synthesis) has been reported on nanostructured Ti compared with nano-smooth Ti.[65] Similarly, Peng et al.[66] showed significantly enhanced EC proliferation and secretion of (prostaglandins) PGI_2 on nano-tubular TiO_2 surfaces formed *via* anodic oxidation Generally speaking most studies strongly indicate enhanced EC function on such surfaces.

Finally with respect to EC response, it is interesting to note that even a simple chemical treatment can lead to nano-rough topographies. For example, Serrano et al.[67] showed improved EC adhesion and proliferation on nano-rough poly(caprolactone) films (compared with untreated samples) synthesized using sodium hydroxide (NaOH) etching. In contrast, Miller et al.[68] revealed decreased EC adhesion and proliferation on poly(lactic/glycolic) acid (PLGA) films with surface nano-topographies (compared with micron-sized) created using NaOH etching.

2.5.5.2 Morphology and Response of Smooth Muscle Cells

Micro- and nano-grated topographies have been shown to influence the response of smooth muscle cells (SMCs) on surfaces, particularly those composed of polymers such as PDMS and poly(methyl methacrylate) (PMMA). The general trend in findings is that micro-grated polymers yield increased alignment, elongation and orientation but with reduced proliferation.[69] For example, Yim et al.[70] showed significant elongation, alignment and reduced proliferation of SMCs on PMMA and PDMS nanopatterned gratings produced by nano-imprint lithography. The same authors also demonstrated that a nanostructured surface favored the polarization of microtubule organizing centers in the direction of the cell alignment axis, whereas in the case of non-structured surfaces, microtubule organizing centers were polarized towards the wound edge.

With regard to nano-pits, it has been shown that there is no substantial influence of Ti_6Al_4V surfaces (produced by treatment with H_2SO_4/H_2O_2) on the growth of SMCs.[71] Similarly, in another study, Nguyen et al.[72] showed an unchanged response in SMC adhesion, an alteration in cell morphology and enhanced cell proliferation for cells grown on 200 nm pit surfaces compared with 20 nm pit surfaces. Interestingly, this study also reported that exposure of SMCs to 200 nm pits induced the expression of various genes involved in cell cycle processes such as DNA replication, cell proliferation and signaling transduction pathways.

Finally, numerous studies have examined the response of SMC on random nano-topographies. Gao et al.[73] studied the effect of SMCs in the presence of serum on nano-rough poly(glycolic acid) (PGA) fibers prepared by surface hydrolysis with NaOH. This work demonstrated a significant increase in the density of SMC on surface-hydrolyzed PGA fibers. Increased bladder cell and vascular SMC adhesion and proliferation have also been reported on nano-structured casts of PLGA, polyurethane (PU) and poly(caprolactone) (PCL) films that were chemically treated with NaOH[67] or HNO_3.[74] In contrast, Peng et al.[66] demonstrated decreased proliferation of SMCs and increased expression of smooth α-actin on nanotubular TiO_2 surfaces.

2.5.5.3 Morphology and Response of Fibroblast Cells

Fibroblast cells (FCs) have been widely studied as a model to investigate the influence of micro- and nano-grating island, pit and post geometries on cellular function. As for the studies on ECs and SMCs outlined above, emphasis has been placed on cellular adhesion, alignment, elongation and contact guidance on a variety of substrate materials with micro-grating topographies.[75,76] For example, Lee et al.[77] studied the influence of micro-grating dimensions on the behavior of FCs cultured on Ti substrates and demonstrated that micro-gratings of Ti with widths of 10 and 30 µm, as well as a depth of 3.5 µm, increased the cell viability, proliferation and up-regulation of fibronectin and integrin genes. In addition, human FCs elongated and aligned better on the micro-grated Ti substrates compared with the smooth Ti surfaces. An analogous result was obtained in work on FC response to grating patterns in the nanometer to micron range on PDMS surfaces.[78] Loesberg et al.[79] showed that a lower threshold in grating depth (35 nm) induced the alignment and orientation of FCs. They also reported that gratings of depths less than 35 nm and widths less than 100 nm resulted in no cell alignment. Similarly, Sun et al.[80] studied the geometrical control of FCs on proton micro-machined 3D PMMA scaffolds.

Many studies have been conducted on FC response to nano-island geometry. Berry et al.[81] demonstrated increased FC adhesion, spreading, morphology and cytoskeleton organization on nylon tubes exhibiting an internal nano-topography generated through polymer de-mixing. Similarly, Dalby et al.[82–85] in a number of works reported a wide range of FC responses to nano-island topographies with heights of 10, 13, 27, 35, 45, 50 and 95 nm that were also created using polymer de-mixing. The 10, 13 and 27 nm islands increased FC adhesion, proliferation, cytoskeletal development and up-regulation of gene expression. Cell adhesion and proliferation on the 95 nm islands were reduced and the cells displayed the most stellate morphologies with poorly formed cytoskeletons. Cells on the 35, 45, and 50 nm island surfaces had the same surface area as cells on flat controls, but with a less developed cytoskeleton. These studies highlight the fact that slight changes in feature height can produce very varied cellular responses. These experiments, however, did not reveal why the cells showed increased or reduced adhesion and growth on the different nano-islands. This type of surface structure created by polymer de-mixing is limited to polymers.

Several studies have looked at the response of FCs to nano-pits. One such study utilized arrays of nano-pits produced by e-beam lithography. The results were increased cell spreading and filapodia interactions on 120 nm pits compared with 75 nm pits, while cells on 35 nm pits had a similar number of filopodia to those grown on control samples.[86]

The effect of nano-post geometries on FC function has been the subject of much study. Green et al.[87] found that posts of 2 and 5 µm heights resulted in increased cell proliferation compared with 10 µm high posts and smooth surfaces. Milner and Siedlecki[88] observed FC adhesion and proliferation on PLA surfaces patterned with 400 nm and 700 nm posts *via* replication molding. Their results demonstrated increased FC adhesion and decreased cell proliferation on surfaces with 400 nm textures compared with 700 nm textures and smooth surfaces. The effect of FC adhesion on polycarbonate and polyetherimide surfaces with micro-post features of varying dimensions, which were generated by laser treatment, demonstrated no cellular orientation with respect to posts. FCs spread and elongate whether in contact with posts or microsmooth materials.[89] Similarly, Dalby et al.[90] demonstrated reduced FC adhesion on PMMA substrates with nano-posts (prepared by colloidal lithography) that were 100 nm in diameter, 160 nm in height and having a pitch of 230 nm. An increase in endocytosis was also noted on nano-pits using clathrin staining indicating that these nano scale features are in the same size range as those features with which the *in vivo* cells interact with. FCs have also been studied on silicon surfaces possessing nano-posts with a height ranging from 50–100 nm and with a pitch of 230 nm that were fabricated by interference lithography and deep reactive-ion etching.[91] Human FCs were found to attach in a similar density on flat control surfaces. However, the cell morphology was more elongated on the nano-posts, an effect noticed for up to seven days in culture.[92]

FC behavior has also been investigated on random nano-rough geometries. It has been shown that decreased fibroblast numbers occur on NaOH treated PLGA and PCL surfaces, as well as nitric acid (HNO_3) treated PU.[93] Cousins et al.[94] have also shown that nano-rough surfaces created with silica nano-particles affect FC morphology, decrease cell adhesion and inhibit cell spreading and thus cell proliferation for periods of up to seven weeks.

In general, stronger alignment, elongation, migration and decreased proliferation of FCs were reported on grated surfaces compared with island, post, pit and random topographies. The studies described here also showed that feature depth, height and surface chemistry have a significant influence on cellular response.

2.5.6 Substrate Rigidity and Elasticity

In contrast with the voluminous amount of research on the role played by surface morphology on cellular response, the rigidity and elasticity of substrates have received relatively little attention. As would be expected, by far the majority of this research has involved polymer-based substrates. The main thrust of this type of research is to use matrices that attempt the mimicking of the properties of natural tissue in terms of stiffness and elasticity. The obvious implication is that such matrices can be employed as scaffolds in implant technology. We outline some examples in this section.

The effects of the flexibility of a substrate on cellular behavior was reported as early as the 1990s.[95] Polyacrylamide gels displaying differential elasticity were fabricated on glass cover slips and characterized by measurement of Young's modulus. Epithelial and fibroblast dells were cultured on collagen-coated polymer films in order to examine mobility and adhesion characteristics, *etc*. This experiment was designed in this fashion to allow polymer flexibility to change without ostensibly changing any surface chemical conditions. Cells were fixed and studied by standard fluorescence, microscopy and immunoblotting protocols. The important result emanating from this work was that cells grown on flexible surfaces showed increased mobility and reduced spreading. This result was ascribed to mechanical effects on the cell cytoskeleton, possibly through influences on receptor association.

A similar strategy of variation of polymer properties was employed in order to examine the effect of substrate mechanical behavior on neurite extension of dorsal route ganglion in 3D culture experiments.[96] Gels with different concentrations of agarose were prepared and characterized in the terms of their mechanical properties by standard rheometry. The rate of neurite extension for the various gel substrates was performed using time-lapse microscopy and it was shown that the higher the gel concentration, the lower the rate of neurite extension. A mathematical model was applied to the results in an attempt to derive a relationship between polymer mechanical properties and ganglion behavior. A significant conclusion was reached that scaffolds for potential neural growth should involve an appreciation of scaffold elasticity.[96]

Collagen-coated polyacrylamide gels were used in a study of the spreading and other properties of aortic smooth muscle cells in a similar approach to that mentioned above.[97] In this case the elasticity of the polymer substrates were determined suing a cantilever-based atomic force microscopy (AFM) method. The elastic modulus could be calculated from a literature procedure. In a second appraisal of gel stiffness a macroscopic tension method was used for comparison with the AFM results. Cell spreading, cell shape and cytoskeletal or focal adhesion assembly and fluorescent intensity was studied by fluorescence microscopy following the usual cell fixing and staining protocols. The SMCs spread significantly higher on 'rigid' collagen-coated glass compared with the gels in agreement with the work outlined above on epithelial and fibroblast cells. (Any role played by the collagen layer in terms of stiffness was ignored.) Cell spreading properties found in this work are reproduced in Figure 2.16. Interestingly, the authors ascribe the peaks observed in the plots to SMC crawling capability associated with cell ligand density. The biphasic phenomena exhibited in the plots is attributed to cell crawling being initially limited at low ligand densities whereby a cell cannot form adequate attachments to pull itself forward or spread. But at high ligand densities, a cell cannot detach from enough ligand to bring its rear forward. In addition to cell spreading the authors also investigated the role of collagen much as was the case of studies described in section 2.5.4.

Finally, it is noteworthy that research involving substrate stiffness has led to direct attempts to compare the compliance of natural tissue with synthesized chemistries. An example of this sort of study is that of Georges et al.[98] on neuron and glial cell growth in mixed cortical cultures. In this case, two types of polymer matrix, fibrin and the more usual polyacrylamide, were investigated. The dynamic shear modulii of both gels were measured *via* spectrometric rheometry and cell interactions with the matrices were examined by standard immunocytochemistry. It was found that, on 'soft' gels, astrocytes do not spread and have disorganized F-actin compared with the cytoskeletons of astrocytes on 'hard' surfaces. Neurons, however, extend long neurites and polymerize actin filaments on both soft and hard gels. Laminin-coated soft gels

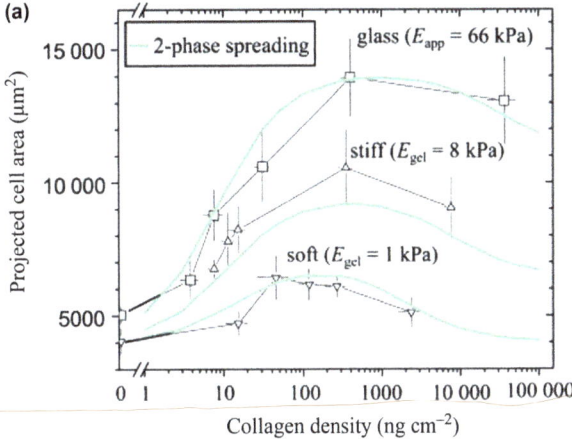
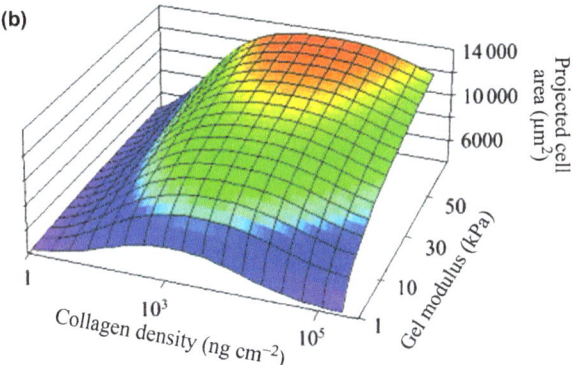

Figure 2.16 Spread cell area as a function of ligand density on soft, stiff, and rigid substrates. (a) The projected cell area was determined 4 h after plating ($n > 10$ per datapoint), giving the indicated average (mean ± standard error). The smooth curves are calculated from a model for two-phase spreading expressed in terms of both E (or E_{app}) and collagen density. Note that cells respond strongly to increasing collagen density on glass and hardly at all on soft gels. (b) Curved surface in three dimensions that fits SMC spreading.
(Reprinted by kind permission from CellPress, Biophysical Society.)

induce attachment and growth of neurons, whereas they suppress astrocyte growth. The number of astrocytes on soft gels is lower than on hard gels, even in the absence of mitotic inhibitors normally used to temper the astrocyte population. The stiffness of materials required for optimal neuronal growth, characterized by an elastic modulus of several hundred Pascals, is in the range measured for intact rat brain. The authors concluded that their results emphasize the potential importance of material substrate stiffness and elasticity as an essential design compound in terms of future biomaterials intended to promote neuronal regeneration across a lesion in the central nervous system.

2.6 Biocompatibility and the Substrate–Blood and Platelet Interaction: A Comment on Long-term Effects

The huge importance of the biochemistry of interaction of cells with substrate surfaces in terms of biocompatibility has figured significantly in the summarizing material presented above. Here we address, as a final comment, the damaging effects that the interaction of biological elements, including cells, suffers as a result of this type of process. The subsequent cascade of events can lead to serious medical problems. Blood interactions with bio-active surfaces are of course ubiquitous in medicine with respect to both intra- and extra-corporeal devices. Included are a host of implantables such as artificial hips and stents placed inside the body and extracorporeal structures such cardio-pulmonary bypass circuitry, oxygenators, and dialysis and hemoperfusion membranes. To fabricate these systems a very wide variety of materials are employed which include many polymers and metals such titanium and steel. A key issue with regard to blood–surface interactions, whatever the substrate, is the possibility of increased thrombogenicity. For the non-biologist this means the increased risk of clot formation in the vasculature. The foreign body can result in the accumulation of circulating platelets at its surface. These in turn are activated, culminating in thrombus formation *via* an aggregation of thrombin and fibrin. Of course there are many possible biological causes for thrombus formation, which is now well-understood through research in cellular and molecular biology. However, the role of foreign substrates is perhaps less well characterized.

With respect to cardiopulmonary systems such as dialysis, hemoperfusion and bypass membranes, there is evidence for a persistent and debilitating effect on the brain function which is manifested as cognitive deficit. It has been shown by Stroobant and Van Nooten[99] that many patients who undergo cardiopulmonary bypass surgery present post-surgical cognitive defects. The mechanism of this process is thought to involve cerebral micro-strokes caused by thrombi release from, for example, perfusion apparatus due to platelet activation and protein coagulation. In a similar vein, it has also been shown that patients who receive continual hemodialysis treatment for renal failure exhibit a progressive decline in cognitive function.[100] Again, at least part of this decline is thought to involve foreign surface instigation of inflammatory effects.

Coronary artery disease can be treated by bypass surgery (and balloon angioplasty). An alternative to this approach was introduced by Sigwart *et al*.[25] in terms of the first coronary stent, a medical device designed to serve as a temporary or permanent mechanical support within the artery. Stents are meshed cylindrical scaffolds, usually constructed from metals or polymers. The stent is typically inserted *via* balloon- or self-expandable technology into the vascular lumen and expanded into contact with the diseased portion of the arterial wall, restoring adequate blood supply to the heart muscle. However, neointimal hyperplasia (NIH) has remained the principal cause of in-stent re-stenosis, which results mainly from over-proliferation of vascular SMCs and production of extracellular matrix. The limitations exhibited by bare metal stents have led to the search for new stent materials, designs, surface treatments and coatings to improve the performance of the device, in particular, attenuation of the rate of in-stent re-stenosis.

Stainless steel is by far the most common currently approved material or stent technology due to its excellent mechanical properties. There are five basic types of stent configuration—coil, helical spiral, woven (braided or knitted), ring (individual or sequential) and cell (closed or opened). The devices differ slightly in strut pattern, width, length, diameter, inter-unit connections, geometry, flexibility, radial strength, radiopacity, surface area coverage, metal content, metal composition and/or their delivery system.[26] These differing characteristics appear to profoundly influence both thrombosis and re-stenosis rates. In order to enhance biocompatibility, therapeutic compounds have been introduced to the device surface in so-called drug eluting stent (DES) technology. Despite this advance a number of problems have become evident in several studies such as the uncontrolled release of drug, lack of clinical evidence with respect to the reduction of NIH, and disadvantageous generation of debris.

Finally, we mention the strategy of re-endothelialization of stent surfaces in order to generate a 'natural' environment.[101,102] This approach represents an interesting adjunct to the usual philosophy for cell–surface studies outlined above in the sense that, in this case, cells on a surface are being employed to enhance biocompatibility. Earlier work involved simple seeding with cells prior to and after implantation, but this approach can lead to cell loss. Recently, the

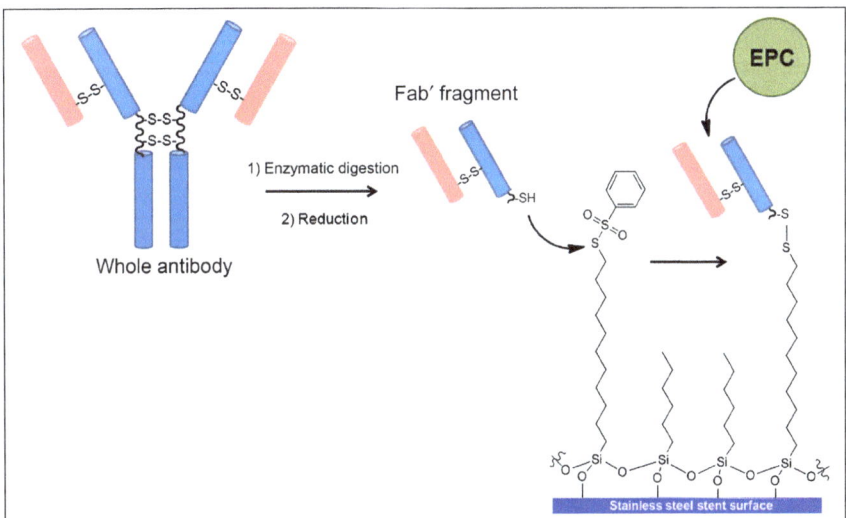

Figure 2.17 Strategy for surface attachment of endothelial progenitor cells *via* oriented Fab fragment of antibody. Fragment is a selective binder for cell surface protein.

notion of capture of circulating progenitor cells by CD34 antibody placed on the device surface was introduced.[103] In our own work, we are studying the possibility that increased biocompatibility of the steel surface can be combined with oriented antibody *via* use of a covertly bound Fab1 fragment.[104] The former concept is designed to reduce protein absorption and the latter the maximization of antibody interaction cell surface receptors. An analogous approach has been employed in biosensor technology for detection purposes, but instead of an antibody for cell capture, a probe is embedded in the antifouling layer. The overall strategy employed to enhance biocompatibility in this case is depicted in Figure 2.17.

References

1. D. J. O'Connor, B. A. Sexton and R. St. C. Smart (eds.), *Surface Analysis Methods in Materials Science*, 2nd ed., Springer, Berlin and London, 2003.
2. J. C. Vickerman and I. S. Gilmore (eds.), *Surface Analysis: The Principal Techniques*, 2nd ed., Wiley, Chichester, 2009.
3. Y. Maréchal, *The Hydrogen Bond and the Water Molecule: The Physics and Chemistry of Water, Aqueous and Bio Media*, Elsevier, Amsterdam and Oxford, 2007.
4. R. J. Clarke, *Adv. Colloid. Interf. Sci.*, 2001, **89–90**, 263.
5. B. Alnerts, D. Bray, J. Lewis, M. Raff, K. Roberts and J. D. Watson, *Molecular Biology of the Cell*, 3rd ed., Garland Publishing, New York, 1994.
6. A. Schapira, *Biochemist*, 2005, **27**, 24.
7. R. O. Hines and K. M. Yamada (eds.), *Extracellular Matrix Biology*, Cold Spring Harbor Laboratory Press, New York, 2012.
8. M. J. Humphries, *Biochem. Soc. Trans.*, 2000, **28**, 311.
9. E. Ruoslahti, *Ann. Rev. Cell Dev. Biol.*, 1996, **12**, 697.
10. Y. Mao and J. E. Schwarzbauer, *Matrix Biol.*, 2005, **24**, 389.
11. J. F. Stein and C. Stoodley, *Neuroscience: An Introduction*, Wiley, Chichester, 2006.
12. G. L. Fain, *Molecular and Cellular Physiology of Neurons*, Harvard University Press, Cambridge, MA, 1999.
13. J. G. Nicholls, *From Neuron to Brain*, Sinauer Associates, Sunderland, MA, 2001.
14. D. A. Lovejoy, *Neuroendocrinolgy: An Integrated Approach*, Wiley, Chichester, 2005.
15. E. Katz (ed.), *Molecular and Supramolecular Information Processing: From Molecular Switches to Logic Systems*, Wiley-VCH, Weinheim, 2012.
16. J. Wu (ed.), *Biomedical Engineering and Cognitive Neuroscience for Healthcare: Interdisciplinary Applications*, IGI Global, Hershey, PA, 2013, 17.
17. R. Carter, *The Human Brain Book*, DK Publishing, New York, 2009.

18. G. M. Shepherd, *Electrophysiology of the Neuron*, Oxford University Press, Oxford, 1994.
19. S. K. Sia and G. M. Whitesides, *Electrophoresis*, 2003, **24**, 3563.
20. J. Fried and L. Yeo, *Biomicrofluidics*, 2010, **4**, 026502.
21. H. Wu, B. Huang and R. N. Zare, *Lab on a Chip*, 2005, **5**, 139.
22. J. H. Park, K. D. Park and Y. H. Bae, *Biomaterials*, 1999, **20**, 943.
23. J. C. McDonald and G. M. Whitesides, *Acc. Chem. Res.*, 2002, **35**, 491.
24. J. N. Lee, X. Jiang, D. Ryan and G. M. Whitesides, *Langmuir*, 2005, **20**, 1168.
25. U. Sigwart, J. Puel, V. Mirkovitch, F. Joffre and L. Kappenberger, *N. Engl. J. Med.*, 1987, **316**, 70.
26. F. Nazneen, G. Herzog, D. W. M. Arrigan, N. Caplice, P. Benvenuto, P. Galbin and M. Thompson, *J. Biomed. Mat. Res., B*, 2012, **100B**, 1989.
27. A. Shahryari, F. Azari, H. Vali and S. Omanovic, *Acta Biomater.*, 2010, **6**, 695.
28. D. Cadosch, E. Chan, O. P. Gautschi, H. P. Simmen and L. Filgueira, *J. Orthop. Res.*, 2009, **27**, 841.
29. P. P. Vicario, Z. Lu, I. Grigorian, Z. Wang and T. Schottman, *J. Biomed. Mat. Res. B*, 2009, **89B**, 114.
30. X. Wang, J. S. Ellis, C.-D. Kan, R.-K. Li and M. Thompson, *Analyst*, 2008, **133**, 85.
31. L.-E. Cheran, S. Cheung, D. Belsham, W. A. MacKay, A. Al Chawaf, J. S. Ellis, D. Lovejoy and M. Thompson, *Analyst*, 2007, **132**, 242.
32. M. Schena (ed.), *DNA Microarrays: A Practical Approach*, Oxford University Press, Oxford, 1999.
33. M. Schena, *Protein Microarrays*, Jones and Bartlett Learning, Boston and London, 2005.
34. D. Mazier, G. Schatten and D. W. Sale, *J. Cell. Biol.*, 1975, **66**, 198.
35. G. Rainaldi, A. Calcabrini and M. T. Santini, *Mater. Sci. Mater. Med*, 1998, **9**, 755.
36. C. Richter, M. Reinhardt, S. Giselbrecht, D. Leisen, V. Trouillet, R. Truckenmüller, A. Blau, C. Ziegler and A. Welle, *Biomed. Microdevices*, 2010, **12**, 787.
37. S. S. Rao, N. Han and J. O. Winter, *J. Biomed. Sci. Polym. Ed.*, 2011, **22**, 611.
38. S. G. Lévesque and M. S. Shoichet, *Biomaterials*, 2006, **27**, 6277.
39. P. Musoke-Zawedde and M. S. Shoichet, *Biomed. Mat.*, 2006, **1**, 162.
40. M. R. Hynd, J. P. Frampton, N. Dowell-Mesfin, J. N. Turner and W. Shain, *J. Neurosci. Methods*, 2007, **162**, 255.
41. L. M. Y. Yu, N. D. Leipzig and M. S. Shoichet, *Mater. Today*, 2008, **11**, 36.
42. G. M. L. Le Saux, *Modified Silicon Surfaces for Controlled Cell Interactions*, PhD thesis, The University of New South Wales, Australia, 2010.
43. S. P. Massia and J. A. Hibbell, *J. Cell. Biol.*, 1991, **15**, 1089.
44. E. A. Cavalcanti-Adam, T. Volberg, A. Micoulet, H. Kessler, B. Gerger and J. P. Spatz, *Biophys. J.*, 2007, **92**, 2964.

45. J. Huang, S. V. Gräter, F. Corbellini, S. Rinckm, E. Bock, R. Kemkemer, H. Kessler, J. Ding and J. P. Spatz, *Nano Lett.*, 2009, **9**, 1111.
46. C. Roberts, C. S. Chen, M. Mrksich, M. Martichonok, D. E. Ingber and G. M. Whitesides, *J. Am. Chem. Soc.*, 1998, **120**, 6548.
47. N. T. Flynn, T. N. T. Tran, M. J. Cima and R. Langer, *Langmuir*, 2003, **19**, 10909.
48. M. R. Linford and C. E. D. Chidsey, *J. Am. Chem. Soc.*, 1993, **115**, 12631.
49. K. A. Kilian, T. Bocking, K. Gaus, M. Gal and J. J. Gooding, *Biomaterials*, 2007, **28**, 3055.
50. G. M. L. Le Saux, A. Magenau, T. Böcking, K. Gaus and J. J. Gooding, *PLoS One*, 2011, **6**, e21869.
51. S. H. Ngalim, A. Magenau, G. M. L. Le Saux, J. J. Gooding and K. Gaus, *J. Oncol.*, 2010, **2010**, 363106.
52. S. Sheikh, J. C.-C. Sheng, C. Blaszykowski and M. Thompson, *Chem. Sci.*, 2010, **1**, 271.
53. S. Sheikh, D. Y. Yang, C. Blaszykowski and M. Thompson, *Chem. Commun.*, 2012, **48**, 1305.
54. J. Norman and T. Desai, *Ann. Biomed. Eng.*, 2006, **34**, 89.
55. A. Tseng and A. Notargiacomo, *J. Nanosci. Nanotechnol.*, 2005, **5**, 683.
56. J. C. Palmaz, A. Benson and E. A. Sprague, *J. Vasc. Interv. Radiol.*, 1999, **10**, 439.
57. J. Lu, M. Rao, N. MacDonald, D. Khang and T. J. Webster, *Acta Biomater.*, 2008, **4**, 192.
58. A. Ranjan and T. J. Webster, *Nanotechnology*, 2009, **20**, 305102.
59. S. Y. Hwang, K. W. Kwon, K. J. Jang, M. C. Park, J. D. Lee and K. Y. Suh, *Anal. Chem.*, 2010, **82**, 3016.
60. C. Bettingerm, Z. Zhang, S. Gerecht, J. T. Boenstein and R. Langer, *Adv. Mater.*, 2008, **20**, 99.
61. S. H. Hsu, C. M. Tang and C. C. Lin, *Biomaterials*, 2004, **25**, 5593.
62. M. J. Dalby, M. O. Riehle, H. Johnstone, S. Affrossman and A. S. G. Curtis, *Biomaterials*, 2002, **23**, 2945.
63. P. Kim, D. H. Kim, B. Kim, S. K. Choi, S. H. Lee, A. Khademhosseini, R. Langer and K. Y. Suh, *Nanotechnology*, 2005, **16**, 2420.
64. J. D. Zawislak, K. W. Kolasinski and B. P. Helmke, *Phys. Status Solidi A*, 2009, **206**, 1356.
65. S. Choudhary, K. M. Haberstroh and T. J. Webster, *Tissue Eng.*, 2007, **13**, 1421.
66. L. Peng, M. L. Eltgroth, T. J. LaTempa, C. A. Grimes and T. A. Desai, *Biomaterials*, 2009, **30**, 1268.
67. M. C. Serrano, M. T. Portoles, M. Vallet-Regi, I. Izquierdo, L. Galletti, J. V. Comas and R. Pagan, *Macromol. Biosci.*, 2005, **5**, 415.
68. D. C. Miller, A. Thapa, K. M. Haberstroh and T. J. Webster, *Biomaterials*, 2004, **25**, 53.
69. G. R. Houtchens, M. D. Foster, T. A. Desai, E. F. Morgan and J. Y. Wong, *J. Biomech.*, 2008, **41**, 762.

70. E. K. F. Yim, R. M. Reano, S. W. Pang, A. F. Yee, C. S. Chen and K. W. Leong, *Biomaterials*, 2005, **26**, 5405.
71. I. Richert, F. Vetrone, J. H. Yi, S. F. Zalzal, J. D. Wuest, F. Rosei and A. Nanc, *Adv. Mater.*, 2008, **20**, 1488.
72. K .T. Nguyen, K. P. Shukla, M. Moctezuma and J. P. Tang, *J. Nanosci. Nanotechnol.*, 2007, **7**, 2823.
73. J. M. Gao, L. Niklason and R. Langer, *J. Biomed. Mater. Res.*, 1998, **42**, 417.
74. A. Thapa, D. C. Miller, T. J. Webster and K. M. Haberstroh, *Biomaterials*, 2003, **24**, 2915.
75. X. F. Walboomers, W. Monaghan, A. S. G. Curtis and J. A. Jansen, *J. Biomed. Mater. Res.*, 1999, **46**, 212.
76. X. F. Walboomers, L. A. Ginsel and J. A. Jansen, *J. Biomed. Mater. Res.*, 2000, **51**, 529.
77. S. W. Lee, S. Y. Kim, I. C. Rhyu, W. Y. Chung, R. Leesungbok and K. W. Lee, *Clin. Oral Implants Res.*, 2009, **20**, 56.
78. S. A. Biela, Y. Su, J. P. Spatz and R. Kemkemer, *Acta Biomater.*, 2009, **5**, 2460.
79. W. A. Loesberg, J. te Riet, F. van Delft, P. Schon, C. G. Figdor, S. Speller, J. van Loon, X. F. Walboomers and J. A. Jansen, *Biomaterials*, 2007, **28**, 3944.
80. F. Sun, D. Casse, J. A. van Kan, R. W Ge and F. Watt, *Tissue Eng.*, 2004, **10**, 267.
81. C. C. Berry, M. J. Dalby, D. McCloy and S. Affrossman, *Biomaterials*, 2005, **26**, 4985.
82. M. J. Dalby, M. O. Riehle, H. H. Johnstone, S. Affrossman and A. S. G. Curtis, *Cell Biol. Int.*, 2004, **28**, 229.
83. M. J. Dalby, M. O. Riehle, H. H. Johnstone, S. Affrossman and A. S. G. Curtis, *J. Biomed. Mater. Res. Part A*, 2003, **67A**, 1025.
84. M. J. Dalby, M. O. Riehle, H. H. Johnstone, S. Affrossman and A. S. G. Curtis, *Tissue Eng.*, 2002, **8**, 1099.
85. M. J. Dalby, D. Giannaras, M. O. Riehle, N. Gadegaard, S. Affrossman and A. S. G. Curtis, *Biomaterials*, 2004, **25**, 77.
86. M. J. Dalby, N. Gadegaard, M. O. Riehle, C. D. W. Wilkinson and A. S. G. Curtis, *Int. J. Biochem. Cell Biol.*, 2004, **36**, 2005.
87. A. M. Green, J. A. Jansen, J. Vanderwaerden and A. F. Vonrecum, *J. Biomed. Mater. Res.*, 1994, **28**, 647.
88. K. R. Milner and C. A. Siedlecki, *J. Biomed. Mater. Res. Part A*, 2007, **82A**, 80.
89. J. A. Hunt, R. L. Williams, S. M. Tavakoliand and S. T. Riches, *J. Mater. Sci. Mater. Med.*, 1995, **6**, 813.
90. M. J. Dalby, C. C. Berry, M. O. Riehle, D. S. Sutherland, H. Agheli and A. S. G. Curtis, *Exp. Cell Res.*, 2004, **295**, 387.
91. C. H. Choi, S. H. Hagvall, B. M. Wu, J. C. Y. Dunn, R. E. Beygui and C. J. Kim, *Biomaterials*, 2007, **28**, 1672.

92. C. H. Choi, S. Heydarkhan-Hagvall, B. M. Wu, J. C. Y. Dunn, R. E. Beygui and C. J. Kim, *J. Biomed. Mater. Res. Part A*, 2009, **89A**, 804.
93. R. J. Vance, D. C. Miller, A. Thapa, K. M. Haberstroh and T. J. Webster, *Biomaterials*, 2004, **25**, 2095.
94. B. G. Cousins, P. J. Doherty, R. L. Williams, J. Fink and M. J. Garvey, *J. Mater. Sci. Mater. Med.*, 2004, **15**, 355.
95. R. J. Pelham, Jr. and Y. Wang, *PNAS*, 1997, **94**, 13661.
96. A. P. Balgude, X. Yu, A. Szymanski and R. V. Bellamkonda, *Biomaterials*, 2001, **22**, 1077.
97. A. Engler, L. Bacakova, C. Newman, A. Hategan, M. Griffin and D. Dische, *Biophys. J.*, 2004, **86**, 617.
98. P. C. Georges, W. J. Miller, D. F. Meaney, E. S. Sawyer and P. A. Janmey, *Biophys. J.*, 2006, **90**, 3012.
99. N. Stroobant and G. Van Nooten, *Chest*, 2005, **127**, 1967.
100. A. Murray, *Adv. Chronic Kidney Dis.*, 2008, **15**, 123.
101. T. Shirota, H. Yasui, H. Shimokawa and T. Matsuda, *Biomaterials*, 2003, **24**, 2295.
102. J. Aoki, P. W. Serruys, H. van Beusekom, A. T. L. Ong, E. P. McFadden, G. Sianos, W. J. van der Giessen, E. Regar, P. J. de Feyter, H. R. Davis, S. Rowland and M. J. B. Kutryk, *J. Am. Coll. Cardiol.*, 2005, **45**, 1574.
103. M. Co, E. Tay, L. Chi Hang, K. Kianm, K. A. Low and J. Lim, *Am. Heart J.*, 2008, **155**, 128.
104. M. Thompson, C. Blaszykowski and P. Benvenuto, US Patent Application No. US 20120076833 A1, 2010.

CHAPTER 3
Electronic Detection Techniques

3.1 A Review of Neuron Field Potentials

Comprised of approximately 10 000 million neurons, the brain is the most important and complicated organ of the body. The basic architecture of each neuron consists of two parts: the soma, the cell body, and the finger like projections that radiate outward from the cell body in all directions, commonly referred to as dendrites. (See Chapter 2 for a further look at neuron structural details.) Simply put, the axon is the output extension of the cell body and the dendrite is the input extension of the cell body. Neurons connect to each other *via* an invisible bridge referred to as a synaptic connection. Several synaptic connections can be made, for example, between two different dendrites, a soma and dendrite, or between small spines that branch off surfaces of dendrites. It is estimated that there are at least 1000 synaptic connections per cell. With such remarkable intricate complexity, cells are able to communicate *via* these electrical and chemical signaling connections.

The question that neuroscientists have long sought to address is how these simple neurological connections allow the brain to function. Though the brain's ability to function is not fully understood, it can be surmised that understanding it requires better appreciation of the formation of the patterns that facilitate cognitive and intelligent response. Understanding the mechanism by which these patterns are generated requires an examination of the inherent forces that drive the system to form emergent complex and transient patterns, allowing higher organic systems such as the brain to process multivariable information. The inner workings of this intricate pattern formation within the brain are investigated through smaller model networks that consist of the essential elemental composition of the larger system. Though these artificial neuronal networks do not afford a means to understand the deeper complexities of the brain, they do offer a platform to study the mechanism of function on a smaller scale.

RSC Detection Science Series No. 1
Sensor Technology in Neuroscience
By Michael Thompson, Larisa-Emilia Cheran and Saman Sadeghi
© M Thompson, L-E Cheran and S Sadeghi 2013
Published by the Royal Society of Chemistry, www.rsc.org

At the heart of pattern formation, is the synchronous activity of population of neurons which in turn rely on a nerve signal generated along an axon or dendrite by the difference in the electric field between the internal environment of the nerve fiber and its external environment. This difference in the electric field is called an electric potential, or field potential, and is established by a difference in the concentrations of various charged ions across in a given membrane.

Stimulation of a single neuron is achieved by using a microelectrode to apply a depolarizing current to a conductive medium in which the neuron resides. In this way, it is possible to remove the charge from the membrane and set up the initiation of a potential impulse of $\sim$mV along the nerve fiber. The neuron membrane, ~ 70 Ångstroms in width, is mainly composed of the phospholipid leaflet, which in itself has a high capacity of ~ 1 uF cm^{-2}, although due to the resistance of ion channels (of the entire membrane), the capacity is about 1000 ohm cm^{-2}. A -70mV potential inside relative to the outside is maintained across the membrane through an ion concentration imbalance of 10:1 sodium, 14:1 chloride, and 1:30 potassium outside of the cell relative to inside of the cell. By controlling the potential and imposing a decrease of 20 mV across the membrane it is possible to cause a sudden change in potential, and thereby create a nerve impulse through changes in the sodium and potassium conductance. These action potentials, also introduced in Chapter 2, translate to extracellular potentials both in the axons and larger potentials of 3–5 mV, which are more spatially confined and observed around the cell body. These extracellular potentials influence neuronal interaction and can be perturbed to create signals that pass through the nerve fiber. Once this electric signal reaches the synaptic connection between two different nerves, the release of special chemicals, neurotransmitters, is stimulated. The neurotransmitters then travel through the synaptic connection to pass along this signal to the neuron on the receiving end.

With careful design it is possible to engineer an *in vitro* environment for neuronal networks and spatially localized electrical stimulation to selectively excite either excitatory or inhibitory neurotransmitters. Ultimately through this fabricated system of various excitatory and inhibitory signal transmissions, additional information can be gathered to better understand the connections between different neurons and their neuronal network signaling patterns.

3.2 Cultured Neurons and Neuro-electronic Interface

As the neuron has characteristic electrochemical properties, it can be viewed as an electrogenic cell. This type of cell generates electronic transduction and facilitates two-way interactions within the cell. Extracellular potentials can be explained with established biophysical principles of neural excitability.[1,2] Sub-threshold action potential currents flow in closed loops which enter at the neuron's current sink. These currents then propagate through the cytosolic medium and leave at the current source to return again to the cell through the sink after travelling in extracellular space.[3] The measured voltages through external electrodes, which can form a two-way interaction with neurons, give rise to spatio-temporal patterns of extracellular voltages that are indicative of the return current. Micro electrode arrays (MEAs) and specialized miniaturized tools, such as complementary metal oxide semiconductor (CMOS) technology, have evolved as emerging tools in neuroscience research to measure these voltages. Spatial distribution of electrodes is important since information in the brain is processed by populations or clusters of neurons rather than by single cells. Measurement of electrical activity in a brain slice (or cultured neuronal network obtained in an *in vitro* extracellular recording technique) enables a further understanding of brain and neuronal network function. Furthermore, measurement of electrical potentials from arrays of electrodes gives physiologically relevant data such as action potential firing rates and long-term measurements,[4] firing patterns of populations of neurons, signal propagation, information processing, memory and learning.[5] In addition to passive measurement of neuronal potentials, active applied experiments through excitation of neurons and observation of changes (resulting from neuronal networks responses to environmental conditions and perturbations) can be used for toxicology studies, drug screening and applications in cellular biosensors.[6]

The neuron is a self-sustaining entity, tolerating the *in vitro* milieu extremely well given the right nutrients and environment. These properties contribute to an ideal experimental setup, where neurons are kept alive over many months on metal electrodes and artificial substrates. Neural networks with controlled orientation, discussed in the next chapter, can be grown on artificial substrates and subjected to electrical and electromagnetic fields, drugs, neurotransmitters and so on to better understand how the cell reacts to various stimuli. Neuron-based electronic devices can measure a range of phenomena including electronic changes in membrane potential, transmission effects, ion channel activity, neuron–neuron connectivity and neuron contact with a high degree of sensitivity. These electronic interfaces can be used as specific biosensors to detect metabolic activity, small molecule interactions, drug effects, toxicity and mutagenicity.

It is fascinating to watch in real time the self-organizing process of a neural network from the seeding on a surface (from isolated neurons) to the formation of a complex. Recent developments in multielectrode arrays, optical imaging and fluorescence microscopy add complementary dimensions to understanding the process developing.

Using a simple charge coupled device (CCD) camera mounted on a phase contrast microscope coupled to a PC (for image processing), it is possible to observe in real time the growth of a neural network.[7] The most intense stage of development happens between day 1 and day 5. After this rapid growth, there is a pronounced decrease in growth rate. By day 6 in culture, most of the neurons develop interconnections and are part of an elaborate network. During the growth process, growth cones connect not just to neighboring cells but also to neurites previously extended from their own cell body with no evidence of self-avoidance. They thereby form close loops, which are thought to have importance in the functional feedback circuits. What happens next is essential and illustrates the amazing plasticity of such networks of living neurons.

After the initial stage of intense neurite formation, the neuronal cell bodies start to aggregate into packed clusters. The clustering of cells is accompanied by absorption of branches, as well as the formation of whole neurites, rearrangements of neurites, and fusion of parallel neurites. The somata can then be observed to migrate along newly formed bundles toward one another. The clustering of neurons is essential in maintaining the synchronous oscillatory activity. This is a characteristic of both the brain circuits and can be observed *in vitro* environments outside the brain.

In an *in vitro* environment, during the first week in culture, the neurons begin to develop spontaneous activity. These activity patterns include complex sequences of action potentials in isolation and in bursts that *in vitro* continue to develop over the course of a month. Underlying these activity changes are morphological changes of the neurons, allowing them to grow elaborate dendritic and axonal arbors to form numerous synaptic connections. The glial cells, if present in the dish, continue to divide and proliferate until limited by contact inhibition or exogenous inhibitors of cell division. Glial cells provide the necessary trophic factors for cultured neurons. There is also evidence that direct contact between neurons and glial cells is crucial for neuronal survival (potentially also playing a role in synaptic processing).

The final morphology of the neuronal networks is generated from two opposing forces. There are benefits and drawbacks to the final morphology. The formation of these networks is comprised of single neurons that grow axons and highly branched dendritic trees to achieve maximum interconnected networks. The benefit of this formation is that it allows information to flow and efficiently travel. The drawback with this formation is that developing extended and vastly branched neurites has a high energetic cost. Hence, the structure of the neuronal network results from interplay between these factors.[8] The process may be viewed as a small-scale illustration of what happens in the brain during the continuous process of rewiring and establishing new connections. What orchestrates such a complex and dynamic process is difficult to imagine. What we know today is that structural plasticity is directed by guiding molecules released by target cells. External stimulation generates new synaptic connections between neurons.

Neuronal activity controls calcium influxes, directing the release of neurotrophic factors, movement of growth cones and synaptic differentiation. The growth cone consists of filopodia and lamellipodia developed by F-actins and microtubules. The cellular spatial organization inside the growth cone changes continuously *via* external cues by the phenomenon of contact guidance.[9,10] The membranes, which are the natural substrate for cells, possess a complex three-dimensional (3D) topography, consisting of nanometer-sized features. These features alter the cellular adhesion, which enables spreading, migration and proliferation. The fact that this organization can occur in a Petri dish (*in vitro* environment outside the brain) demonstrates that the presence of a distant regulatory source is not necessary and that self-organizing abilities are embedded in each growing cell. This explains the remarkable fluidity and plasticity of the brain structure on a macroscopic scale.

Figure 3.1 Examples of topographical features on silicon and plastic surfaces. (A) Silicon nanoscale roughness (scale bar = 20 nm). (B) Silicon nanoscale pillars (scale bar = 1000 nm). (C) Grooves on poly(dimethylsiloxane) (PDMS) (scale bar = 5 nm).[15] (Reprinted by kind permission of Wiley.)

Several techniques have been developed to direct the growth of neurons and their network structure. Designed *in vitro* network platforms can employ topographical and chemical modification of surfaces (Figure 3.1). In large part these techniques try to mimic or amplify the natural topographical effects of the cell's base membranes and to artificially create a designed substrate with nanostructures that stimulate directed growth of neurons. Such modifications illustrated in Figure 3.1 are hypothesized to involve the creation of submicrometer features by careful design such as laser etching, photolithography and other etching techniques or by applying bulk surface modifications such as roughening of the surface.[11,12] Alternative methods employed separately and in addition to topographical changes can involve changing the chemical composition of a surface by binding cell adhesive molecules to promote directed growth.[13] In order to mimic the spatial structure of the brain, 3D matrices consisting of polyethylene glycol (PEG) hydrogel have been employed.[14] In these platforms the hydrogel is supplemented with poly-L-lysine (PLL), which has proved not to be cytotoxic and allows cell culture growth. Such systems can be useful for analyzing the cell interactions *in vitro* in a more natural 3D system.

Glial cells play an essential role in supporting and nourishing the neurons. They provide natural physical cues to migrating neurons. Thus to gain further insight into the study of neurobiology, it is important to consider carefully the contribution of glial cells.[16,17] Figure 3.2 shows the complicated protocol used to obtain a low density dissociated cell culture of hippocampal neurons from embryonic rats or mice. The neurons are cultured on PLL-treated coverslips, which are suspended above an astrocyte feeder layer and maintained in serum free medium. When cultured according to this protocol, hippocampal neurons are appropriately polarized and develop extensive axonal and dendritic arbors to form numerous, functional synaptic connections with one another. Hippocampal cultures have been widely used to visualize the subcellular localization of endogenous or expressed proteins. This allows for imaging of protein trafficking and the defining of molecular mechanisms underlying the development of neuronal polarity, dendritic growth and synapse formation.[18] Protocols such as these may be adapted to generate growth of neurons from

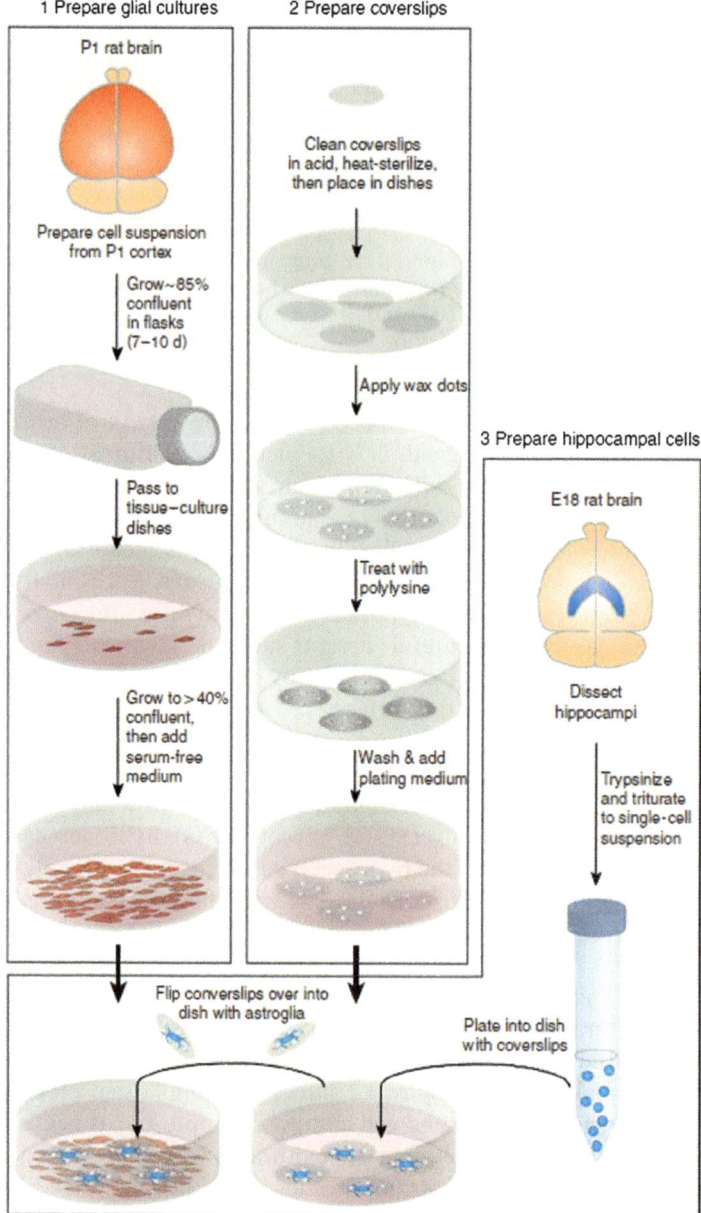

Figure 3.2 Protocol to obtain a dissociated culture of neurons from embryonic cells reprinted.[18]
(Reprinted by kind with permission of the Nature Publishing Group.)

primary cultures on the surface of electrical and optical microarray detecting platforms, allowing for added dimensions of exploring the structural and functional plasticity of neurons.

Electronic Detection Techniques 93

3.3 Charge Transfer and the Interface

The cell contains inherent specific receptors that respond to external stimuli, allowing it to intrinsically behave as a 'sensing' component. The cell's 'sensing' abilities enables interfacing molecular electronics with neurons. In addition, the transduction cascades that are instigated by the initial receptor–ligand binding act as signal amplifiers of the binding event, allowing electronic detection of specific signaling pathways to discern particular interactions taking place at the membrane level.

With advances in the design and fabrication of microelectronic circuits, a perfect dimensional match between electrodes and neurons is generated. Figure 3.3 illustrates the growth of a neuron over a transistor. At the submolecular level, the ionic conduction in the cellular matrix is transferred to electronic conduction in the sensing device. Ions and electrons differ greatly in terms of their mobility and their actual charge transfer mechanism is still not well understood. *In vitro* detection of neural signals requires a stable and carefully designed interface that is capable of optimizing the transduction of action potentials, membrane depolarization, ionic flux and biochemical signals into a quantifiable and measurable electrical signal.

The characterization of the neuron–electrode interface through an electrical model has been reviewed by Rutten[19] and is characterized by three

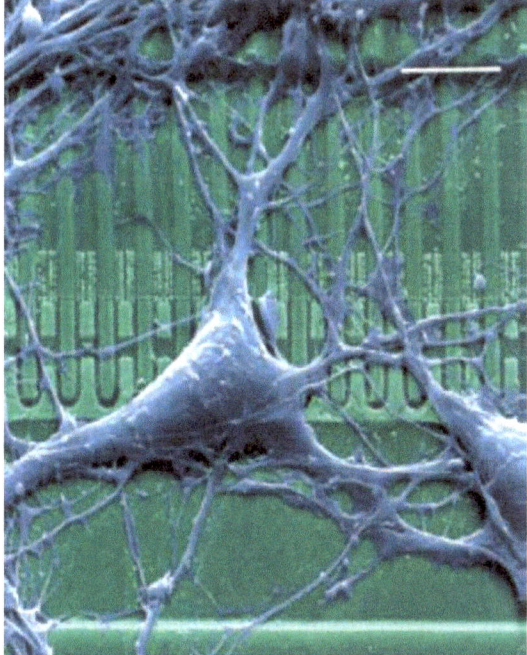

Figure 3.3 Image of a neuron grown on an integrated field effect transistor, and this illustrates the dimensional match between neurons and microelectronics.[20] (Reprinted by kind permission of Elsevier Science.)

components: the neuron, the microelectrode, and the medium in between, which for selective stimulation requires close contact of the electrode with the cell. In an *in vitro* system (a culture dish) a dissociated neuronal cell adheres to a microelectrode array substrate, and in this occurrence, the gap between the cell and the electrode may be very narrow at between 30 and 300 nm.

Electrodes can be designed to exhibit selectivity to discriminately monitor a neuronal process or stimulate a specific element. Examples of this selective stimulation are seen in nerves, fascicles, axons and somata. Thicker neuronal fibers with larger node of Ranvier spacing yield faster conduction as the resistance to the flow of electrical current is inversely proportional to the cross-sectional area of a conductor. Therefore, these fibers with a larger node of Ranvier spacing can be stimulated at a lower current than thinner fibers. For this reason, electrodes with smaller than 10 μm can achieve fiber selectivity by physical proximity to the fiber's node of Ranvier.

The cell–electrode interface dictates the shape, size and position of the electrode with respect to a neuron. The exposure of the electrode to the cell, specific part of the neuron or extracellular medium determines the response of the electrode and even the localization of the electrode's sensing area. Smaller electrodes can be completely covered by the cell soma in which case the electrode interface will only be exposed to the protein membrane and the chemical monolayer of the glycocalix (responsible for the adhesion of the cell to the electrode). The response of the electrode when it completely covered by the soma is simpler to model, as it consists of a parallel resistance and capacitance, and is more localized, In comparison, larger or non-aligned electrodes, which may cover the entire nerve or extend beyond the cell body thereby exposing themselves to the extracellular medium, are more complicated to model.

To optimize microelectrode arrays, there is a correlation between spatial resolution of effected neurons to the magnitude of current that flows through the stimulating electrode. In this way using microelectrodes and passing low currents through optimizes spatial resolution. This is another reason for employing microelectrode arrays. Cultured neurons under low currents and potentials are able to increase the spatial resolution of the interface. Ultrafine electrodes placed close the cell fibers can control and monitor individual neuronal fibers, and potentially empower complete control of the experimental platforms of low density neuronal cultures. From a practical standpoint, however, careful selection of electrode positions at statistically significant locations and avoidance of overlap in the electrode's sensing area are the most experimentally sound methods to achieve selectivity and resolution.[21]

The composition of the electrode material and the charge transfer kinetics at the metal electrolyte interface are also important factors in the stimulation process of neurons. The electrode surface involves flow of capacitive and faradaic currents, as well charge diffusion and migration to the electrode, which are inherently complex processes. To lessen the nonlinear effects of charge transfer and recovery at the interface, it is important to allow enough time between current and potential pulse stimulations to keep the direct current at a constant magnitude, which can allow for full recovery and reestablishment of charge balance.[22] Important factors to consider in the experimental setup are a full understanding of electrode–electrolyte processes and mechanisms, and the mitigation or elimination of currents and mechanisms that are independent of the neuronal interface.

3.4 Field Effect Transistors as Neurotransducers

Decades ago, neuroscientists had limited choices in experimental setup, and typically incorporated a glass micropipette for intracellular recording and a metal wire for extracellular recording. Since then a multitude of sophisticated electrode systems have been developed, incorporating available microfabricating technologies which were initially developed for microelectronic manufacturing. Now the needle or the glass pipette is equipped with onboard electronics (processing chip embedded in the miniaturized device) turning it into an active neural device. In addition, recent technologies are using photolithographic patterning and deposition of thin-film layers on glass, silicon or plastic flexible substrates.

Semiconductor technology employs high density 3D lithography for informing transistors and performing high speed processing. In recent research, it has been shown that this very same technology can be employed to create microelectrodes. When the two technologies are integrated, an improvement in spatial resolution and sensitivity for signal detection from neuronal cultures is illustrated. For example there are dimensional structures that can be built over processing chips using CMOS gate-array technology, as illustrated in Figure 3.4. Silicon microstructures, developed using the same modern microelectronic technology that yielded integrated circuits used in digital logic circuits and microprocessors, can also be employed in the direct interfacing of neuron cells *in vitro* systems.[20,23,24] Neurons can be grown on miniaturized field-effect transistors (FETs) or complementary metal-oxide-semiconductor. Figure 3.4 illustrates the electrical measurement of extracellular activity with a metal electrode and integration of CMOS devices for signal conditioning.

CMOS structures have been designed to continually monitor the changes in the electrical potentials of the cell.[25] The coupling to the cell membrane is achieved capacitively through the electrical polarization of the silicon dioxide layer covering the silicon device.[26] This layer provides a good electrochemical barrier for ionic species, but there is also a lowering of the quality of the recorded electrical signal. At the quantum level, the electrical activity of the neurons affects the electronic band structure of the silicon and is thus transduced into a measurable electrical signal. When the neuron cells are stimulated with an exterior stimulus, the transistor structure provides the electric field, which polarizes the cell membrane resulting in changes to the membrane potential, and the conformation of membrane proteins and the voltage-gated ion channels. The interface also contains the protein molecules (integrins, glicocalix) that are the first to be deposited on the substrate when cell adhesion is initiated. This barrier shields the electric field and suppresses the direct polarization of the membrane. However, the capacitive and ionic

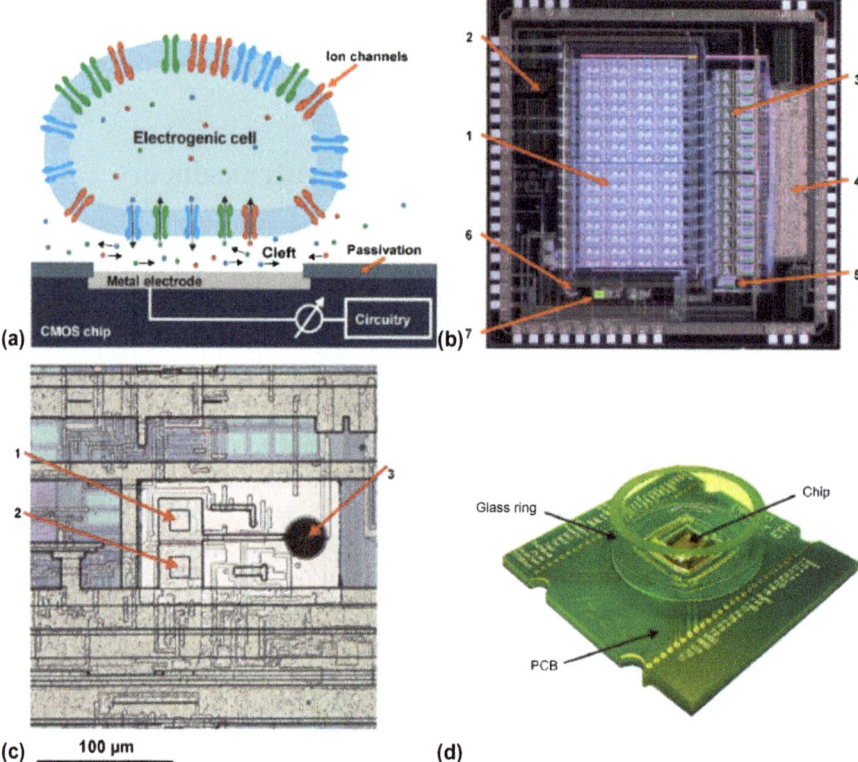

Figure 3.4 (a) Outline of mechanism for extracellular measurement with metal electrodes illustrating the movement of ions in the electrode vicinity to generate a voltage that is detected by the microelectrode. (b) Micrograph of a 6.5 mm^2 CMOS chip encompassing 128 microelectrodes and electronic conditioning circuitry and analog-to-digital conversion electronics for each electrode. (c) Monograph of electrode/cell interface unit with (1) stimulation contacts, (2) recording contacts and (3) platinum black electrode for both stimulation and recording. (d) Representation of packaged chip with an electrode array, integrated CMOS circuitry and printed circuit board (PCB) interconnect, electrodes are housed inside a glass ring used to place cells.[24] (Reprinted by kind permission of Elsevier.)

currents are strong enough to overcome this electrochemical barrier and still provide an accurate reflection of the cellular activity.

Figure 3.5 shows the results of a neuroplasticity experiment in which the regrowth of axons and the formation of new synapses were studied.[20] The recorded potential for an intact Aplysia neuron grown on a CMOS chip is shown in Figure 3.5(A). It can be seen that the potential drastically changes after cutting away the axon, as evidenced in Figure 3.5(B). The potential recovers following the regrowth of neurites from the cut tip, as illustrated in Figure 3.5(C). This experiment demonstrates that the shape and amplitude of field potentials are not only proportional to the intracellular voltage, but also to the anatomy of a particular neuron as well as its ramification pattern (which

Electronic Detection Techniques 97

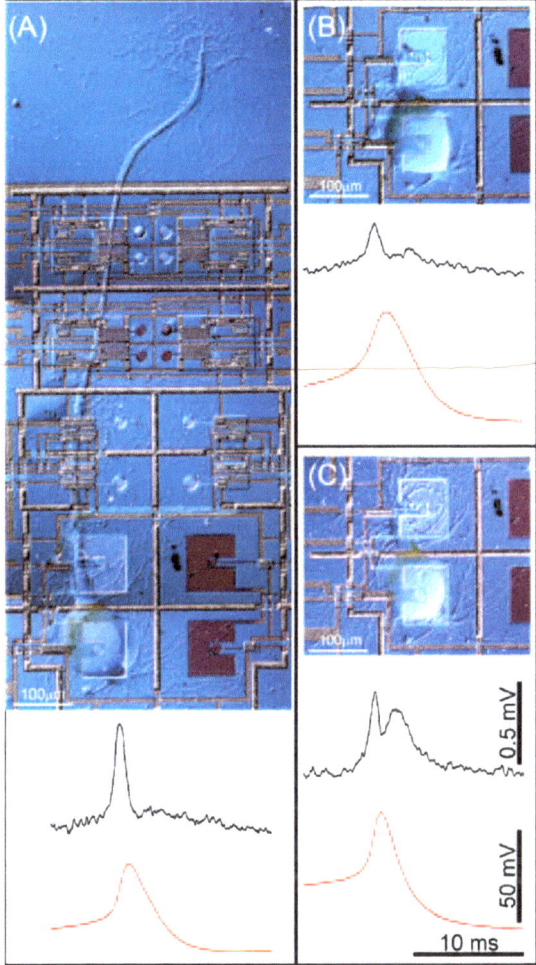

Figure 3.5 (A) Recorded potentials from a neuron cultured on a CMOS chip. (B) After axotomy. (C) 24 h later neurites are regrown from the cut tip.[20] (Reprinted by kind permission of Elsevier Science.)

generates a particular field potential). Such experiments can used to investigate the neuron regeneration process as well as its plasticity. Additionally, this experimental setup may help to evaluate the most optimal use of stem cells in the repopulation or replacement of brain cells, which have been devastated by injury or illness (*e.g.* stroke, Alzheimer's disease, Parkinson's disease, epilepsy, stress injury, leukemia or depression).

3.5 Microelectrode Array Structures

Some of the first *in vitro* studies with neurons were performed using glass micropipette electrodes.[27] As each electrode must be held and tediously positioned by a bulky mechanical micromanipulator, it is very difficult to record or stimulate more than a couple cells at a time. Despite this limitation, neurophysiologists have learned much about single-cell properties, ion channels, pharmacology, and synaptic plasticity in *in vitro* systems. The drawback with these approaches is that many of the collective properties of neuronal networks have been missed, and more importantly, these techniques are invasive, meaning they damage the cells, making long-term monitoring not possible.

Multielectrode arrays are non-invasive, allowing for simultaneous recording and stimulation of neurons. These wired Petri dishes referred to as MEAs, enable the study of distributed patterns of electrical activity in cultured networks *via* non-invasive extracellular electrodes built into the substrate. These electrodes can also be used to stimulate neurons extracellularly and non-destructively, allowing a long-term two-way connection between a cultured neuronal network and a computer.

Thomas *et al.*[28] first described multielectrode arrays for monitoring activity in electrically excitable cells in 1972. A miniature microelectrode array was used to monitor the bioelectric activity of cultured cells. They recorded field potentials from spontaneously contracting sheets of cultured chick cardiac myocytes but could not record activity from single cells. A few years later, Gross (1979)[29] and Pine (1980)[30] independently developed arrays for chronic cell recording and electrical stimulation of cultured neuronal networks.

Until recently, custom-made MEAs, hardware and software were created by each of the laboratories involved in MEA research. Now MEA technology is accessible to all. Complete MEA systems capable of recording from at least 60 electrodes are produced by Multichannel Systems of Germany (MEA60 1), and Panasonic of Japan ('MED System' 2). Guenter Gross (University of North Texas) supplies MEAs that can be used with multielectrode processing hardware and software made by Plexon Inc., USA, Multi Channel Systems GmbH, Germany, and Panasonic Inc., Japan. Only very recently has computer and data storage technology made it feasible to record continuously 60 electrodes at sampling rates over 20 kHz per channel. As technologic advances are made in microfabrication, computer speed and data analysis we will continue to see rapid improvements in the capabilities of commercial MEA systems. The *in vivo* multielectrode probe community is also helping to advance the state-of-the-art, since they share many of the same hardware and data analysis problems with the *in vitro* community.

MEAs usually consist of a number of cell-sized electrodes (10–100 μm) arrayed across the bottom of a cell culture dish. The substrate is usually glass with leads made of gold or the transparent conductor indium tin oxide (ITO) that carry signals from electrodes to external electronics, and carry stimuli to the electrodes. The electrodes must be biocompatible, durable and possess low electrode impedance (less than 500 kilo ohm at 1 kHz) to allow the detection of small cellular signals, which are usually situated in the 10 to 100 microvolt range. With these electrodes it is important to stimulate the neuronal cultures efficiently without inducing a double layer charge build-up and generating overpotentials at the electrode site, which can affect the behavior of neuronal cultures. In order to study the electrode interface, it is essential that minimal impedance is maintained. Existence of high impedance gives rise to high capacitance build-ups and unwanted potentials across the interface.

There are a variety of mechanisms to optimize low impedance. Porous platinum black electroplated while sonicating or sputtering iridium oxide or titanium nitride are acceptable solutions to lower the electrode impedance. Over the electrodes, biocompatible insulating layers are distributed such as polyimide, silicon nitrate or oxide to prevent electrical shorting within the cell growth solution. Cell adhesion is encouraged by depositing a laminin or fibronectin cellular matrix. Alternatively, the fabrication can take place in silicon, using the available microelectronic technologies with capacitive recording and stimulation. To position neurons firmly in place, wells can be used to hold them. The wells can be etched in silicon. However, the neurons often tend to escape from wells, especially if glial cells are nearby to help them attach and move over the margins of the well.

There are ways to optimize CMOS technologies to improve connections. The ever increasing number of electrodes and the quantity of acquired data requires faster processing and better multiplexing of data from arrays of electrodes. The advantages of modern CMOS technologies are related to high connectivity. In CMOS, chip multiplexing is used to enable electrodes with a high spatial and temporal resolution to develop connections with ease through the on-chip logic obtained from the computer systems through digital interfaces and user-friendly software. Extreme miniaturization allows the fabrication of 16 384 electrodes on 6×6 mm chips.[31] However, direct integration of these electrodes with onboard processing is not trivial. Some of the state-of-the-art devices incorporate 128-electrode (addressable) devices mounted on a processing chip (CMOS gate-array mixed-mode technology) (Figure 3.6).[32,33]

Recent advancements in planar microelectrode arrays (PEAs) in the form of active CMOS-integrated electrode arrays[34] are providing new perspectives for the development of high-performance microelectrode arrays. This technology is enhancing the spatial resolution and signal-to-noise ratio of MEAs by incorporating larger numbers of electrodes and the integration of an amplification circuit. High-resolution active pixel-sensor based microelectrode arrays (APS-MEA) exploit the APS technology, which is used in CMOS cameras for fluorescence and light-sensitive measurements.

These interfacing platforms offer new tools for studying the dynamic features of large assemblies of neurons and thus offer promising new perspectives for both fundamental and applied neurophysiology. Neuron cells functioning as a component of biosensing devices can be employed in a number of applications. Hybrid robots, controlled by an array of neurons extracted from the motor cortex of animal embryos and deposited on a glass substrate embedded with electrodes, can direct unmanned aircraft or manage other dangerous situations where the risk to an operator is great. These neural networks, termed live computational devices, are formed by enzymatically extracting the neurons from mature embryos and growing them until they interconnect and establish

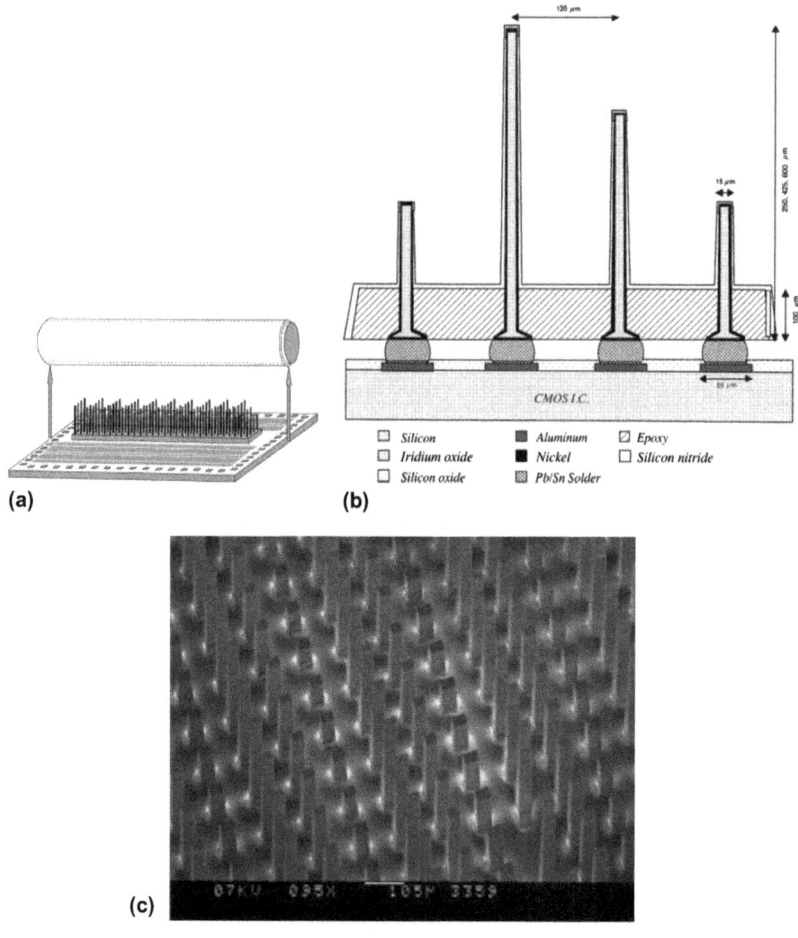

Figure 3.6 (a) a representation of a 3D 128 silicon-glass electrode array. The electrodes are mounted on a CMOS chip with embedded signal processing. The microelectrode needle electrodes are 600, 425 and 250 μm in length, with 120 μm spacing and 15 μm tip. (b) A close-up of the details and composition of the electrode array. (c) An image of etched silicon electrodes of various lengths in a glass matrix.[19]
(Reprinted by kind permission of Annual Reviews).

dendritic connections. An array of 25 000 neurons can convert signals that indicate whether a simulated plane is experiencing stable or volatile weather conditions.[35] Signals related to conditions such as straight flight or tilted axis are transmitted for flight path correction and then retransmitted to the airplane's control system.

Extracellular signals recorded from nerve cell cultures are very sensitive to neuroactive compounds—water-soluble molecules that can have direct metabolic effects. This in turn, may increase, decrease or stop activity (cAMP, cyanide) and cause specific synaptic effects (neurotransmitters and neuromodulators such as strychnine).[36] Additionally, these compounds can stop the propagation of action potentials. For example, tetrodotoxin (TTX) blocks Na^+ channels and oubain causes a similar effect on the sodium pump, preventing the maintenance of the membrane potential. Additionally, there are generic membrane effects mediated by non-synaptic Ca^{2+} or K^+ channels. Both neural biosensor systems presented above (cells cultured on FETs or microelectrodes) are able to detect the effects of such neurochemicals through the changes in membrane potential during an action potential that directly effects the gate of the FET or the electrical capacity of a microelectrode.

Interestingly in use, the built-in sensitivity of neurons enables identification of neurotoxins, such as the Botulinum toxin (a potent neurotoxin produced by toxigenic strains of *Clostridium botulinum*). The Botulinum toxin poses a significant health threat to the public since it can be generated in any insufficiently processed food, or even used deliberately to poison food supplies. As a result, rapid and reliable detection of such a toxin is necessary to reduce the risks of food contamination. A biosensor employing living neural cultures grown *in vitro* on microelectrode arrays can detect such neurotoxin threats with high specificity.

The experimental setup to develop such sensitivity in neurons is discussed further. First, the neurons are grown in a culture dish embedded with a grid of electrodes on its surface. This enables the extracellular recording of action potentials generated by the neural cells. After applying a toxin to the media bath (surrounding the mature neurons), changes are observed. These changes consist of both spontaneous spikes and bursts of activity, which differ relative to control cultures. The application of the toxin also induces a unique oscillatory behavior within each burst that is more reminiscent of early developmental activity patterns in neurons than the behavior of mature cultures as was used in this experiment.[36,37] Once a mature culture is developed above the electrode further research can be conducted.

MEAs are used in frontier stem cell research. For example, MEAs are being used pharmaceutically to develop new novel methods to test active compounds (hypothesized to have drug potential). Researchers have explored the temporal development and pharmacological modulation of the activity in neural networks derived from embryonic stem cells by simultaneously monitoring the *in vitro* electrical activity exhibited by the entire network of neurons over several months.[38] After 5–6 weeks in culture, oscillating and synchronous activity was accompanied by an increase of presynaptic vesicles. The MEA neurochip consisted of 60 planar Ti/TiN microelectrodes of 30 μm diameter situated 200 μm apart. The spike and burst detection software examined the effects of the sodium channel blocker tetrodotoxin, synaptic-acting agonists γ-aminobutyric acid (GABA), *N*-methyl-D-aspartate (NMDA) and combinations with the latter respective antagonists bicuculline and 2-amino-5-phosphonovalerate (APV) by calculating the changes in spike activity compared with baseline activity in the absence of the drug. With this research, the MEA method is evidenced to be a powerful tool with which to investigate pharmaceutically active compounds. Furthermore, it presents a novel approach to the treatment of neurodegenerative diseases by the substitution of lost neurons *via* cell therapy.

Many challenges remain that must be overcome with respect to the detection strategies presented thus far for MEAs. For instance, the low signal-to-noise ratio due to the resistance between the neurons and the FET or microelectrode requires improvement in terms of acquisition and processing of multiple noisy signals. This is thought to be possibly remedied through the use of low-noise transistors and CMOS technology.[39,40] With regard to optimizing neuronelectrode coupling, patterning with aminosilanes or adhesion proteins such as laminin or fibronectin can be utilized. Another challenge to consider is the fact that the number of total active microelectrodes is usually low. Fabrication of smaller electrodes will only worsen the signal-to-noise ratio. Alternative technologies such as CMOS and APS-MEAs that increase the number of measuring sites are currently under development. The data obtained from all the electrodes are usually overwhelmingly large and difficult to analyze. However, computational approaches for pattern recognition in neural response and data mining strategies can be utilized to enhance the high specificity of neural receptor interaction without reducing sensitivity.[41]

New bioinformatics and system biology methods to interpret, classify and link the measured data to neurophysiologic responses are also being explored. As single neurons possess the intrinsic ability to resonate and oscillate at particular frequencies, probing their activity at cellular levels using transistor and microelectrode arrays may reveal the mechanisms responsible for patterns connected to neural signaling.

3.6 Microelectronic Interfaces for *In Vitro* Study of Neurons

MEAs can be designed in a variety of ways to achieve specific experimental aims or to enable and assist in a particular experimental aim. One example is the linear multielectrode, which is a class of one-dimensional array, where an electrode site is mounted on a needle or incorporated in a glass or silicon tip-shaped carrier. The needle can be a hollow metal shaft in which a side-window perforation houses the tips of a number of leads threaded through the shaft. Lithographic patterning and deposition of thin-film metal can also be used to design electrode sites onto glass, silicon, or polyimide carriers.

Another example is a two-dimensional (2D) array. The 2D array consists of electrodes, which may be designed as above, with the tips arranged in the same plane (Figure 3.7). These electrodes can be configured and manufactured as a simple bundle of wires or they may be galvanically or lithographically grown needles.

A third example is the 3D structure of microelectrode arrays. This structure can be achieved if the plane of the arrays is not constant or the electrodes extend to different heights above the substrate (Figure 3.8) and are often employed when spatial selectivity of interaction with neuronal cultures requires height resolutions.[42,43]

Electronic Detection Techniques 103

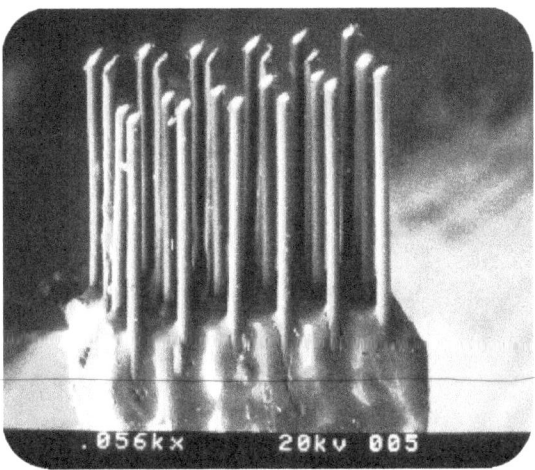

Figure 3.7 Micrograph of 2D wire electrode array with 120 μm spacind.[42]
(Reprinted by kind permission of the IEEE.)

Figure 3.8 3D conductive doped silicon electrode array. Each electrode is insulated with silicon nitride and the exposed tip is platinum plated with a 0.005 mm^2 area. Electrode lengths range from 0.5 to 1.5 mm with 400 μm spacing.[43]
(Reprinted by kind permission of the American Physiological Society.)

Spontaneous activity and induced action potentials can be detected with MEAs using the ion flux associated with channel receptors in the neural membrane. The principle measurement is based on detecting the electrical capacity measured between a chemically modified microelectrode and the neuron axon using ac-coupled amplifiers with high input impedance. Figure 3.9 shows a commercially available microelectrode array with planar 50 μm^2 electrodes.[41] The whole configuration is immersed in a special measurement chamber, which may contain reference electrodes, counter electrodes and/or ion-selective electrodes (for pH or oxygen, *etc.*) which are connected to multiple

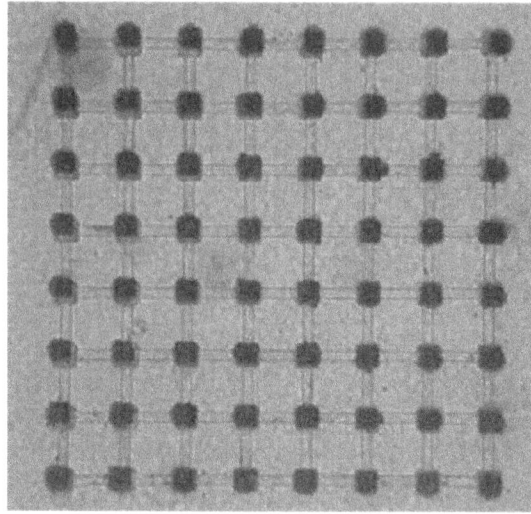

Figure 3.9 A microelectrode array showing planar 50 μm² electrodes (Panasonic Medprobe).

high resolution recording channels, operational amplifiers, filters and long-term data acquisition instrumentation. The experiments also employ electrical stimulation systems, real-time signal visualization and optical surveillance of the cell development using a TV camera connected to an optical microscope positioned over the array. Parameters of interest such as the time of a burst occurrence, the time duration between the first and last spikes, the interval between bursts, changes in synchronization, and the burst amplitude are all recorded and subjected to subsequent analysis.

The analysis of the spatio-temporal patterns in the activity of the neural network provides an insight into the internal dynamics of the neural networks, as well as the relation of synchronization changes to the action of neurotransmitters and their blockers.[6] Figure 3.10 shows a hybrid neuron–MEA interface integrated within a system for neuroscience research aimed towards the understanding of memory formation and application to artificial intelligence.[44]

Hybrid biosensing systems require both the attachment of neurons to a substrate surface and the complete preservation of the neuron's biological activity, viability and functionality upon attachment. In contrast to the patch clamp technique, such systems are not invasive and are capable of monitoring the changes in cell metabolism as a result of input stimuli. The non-biological surface has to be sterile and hospitable to support a neuron or cell. In order to achieve this, the surface is often modified with organic molecules recognizable by the cultured cells. Cell attachment can take place by physisorption, chemisorption or by trapping the neurons in polymeric matrices. Care must be exercised to prevent the creation of unnecessary diffusion barriers that

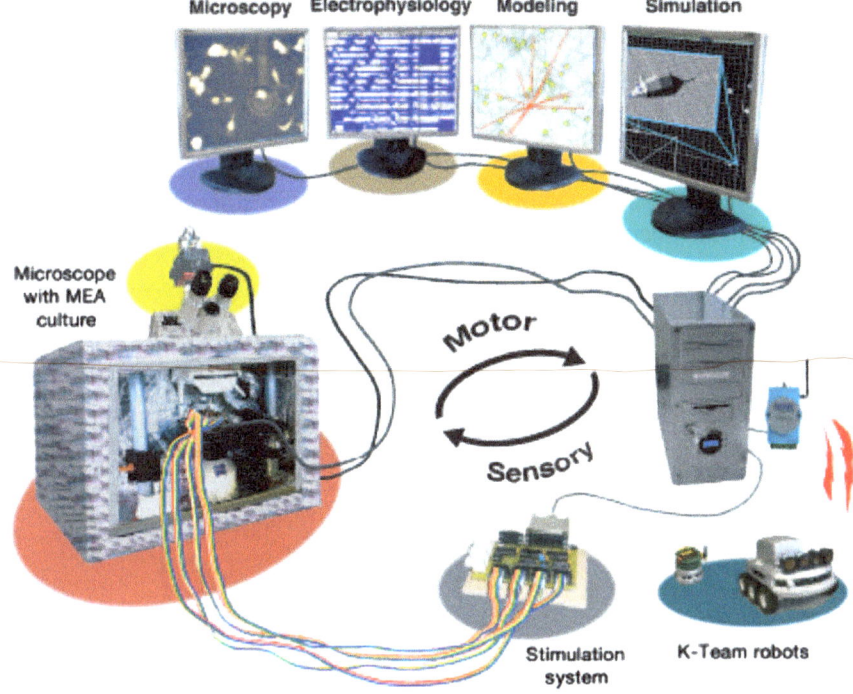

Figure 3.10 A living neural network is grown on a microelectrode array. The activity is recorded, processed in real time and then used to control a robot. (Reprinted by kind permission of Springer Verlag.)

hinder the transport between the neuron membrane and the working electrodes.

3.7 Fabrication

The function of a neuronal network is significantly affected by the neurons' structure and environment, as well as the architecture and organization of the neuronal network. Fabrication of electronic interfaces (to stimulate and detect neuronal activity) is an opportunity to influence and impose an artificial design on neuronal networks to engineer their function. The physical constraints of the *in vitro* environment can direct the growth of neuronal networks and the interconnection. Furthermore, placement of electrodes in strategic locations (in the frame of the platform) assists in stimulation of neurons at specific and critical locations within the network to enhance specific applications. This placement thereby allows for observations of the activity of the system at targeted sites. Fabrication of *in vitro* neuronal network platforms and the placement of electrodes have far reaching implications in the outcome of experiments and, therefore, demand careful consideration of the cells' physiology as well as their interconnection in the design stage.

3.7.1 Regeneration Sieves and Cone-ingrowth Electrodes

When nerve regeneration is envisaged with the intention to regain lost function, two-way interfaces are inserted transversally into the severed nerve. The nerve fascicules slowly grow through the orifices provided. At the same time, the fascicules remain in contact with the electrodes constructed around the respective metalized holes and are strongly fixed to the device in place, which cannot move. This subsequently causes the development of scar tissue because of mechanical micro displacements. Figure 3.11 shows how a planar

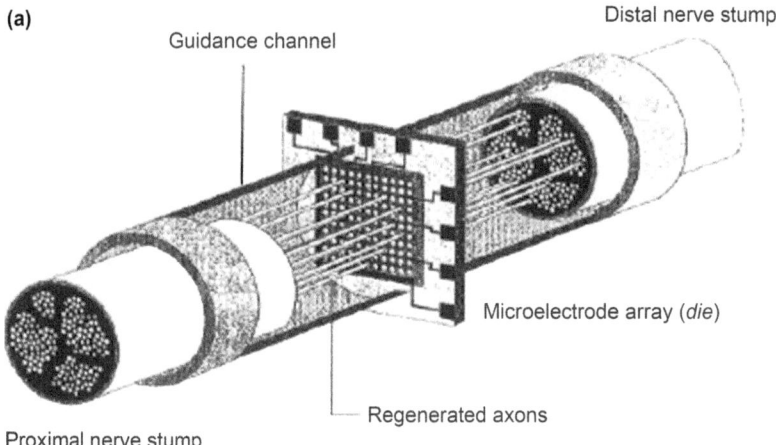

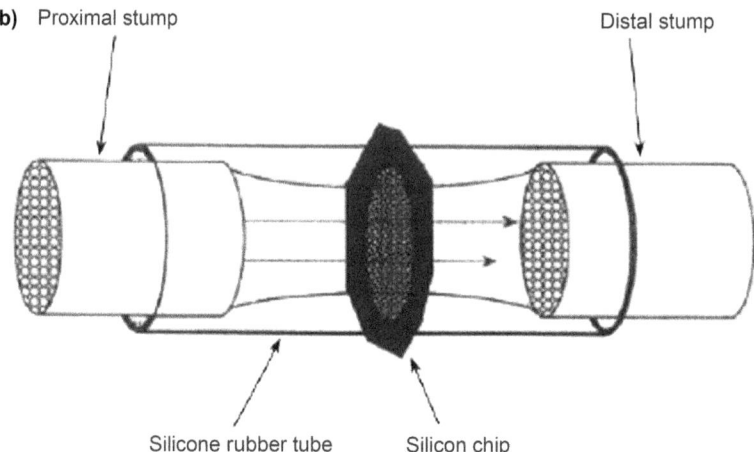

Figure 3.11 (a) Schematic of a designed silver array neural interface implanted to an intersected nerve. (b) Schematic of silicon chamber with a chip bridging the 4 mm gap between the proximal and distal stumps of rat sciatic nerve.[45,46]
(Reprinted by kind permission of Elsevier BV.)

microelectrode sieve is inserted between the two cut ends of a nerve.[45,46] The caveats to this design is the difficulty in placing the sieve near Ranvier nodes, which limits the selectivity of recording or stimulation and the crosstalk between the electrodes of the sieve since they cannot be perfectly insulated from each other. Despite apparent nerve regeneration, the force in the corresponding muscle declines by 40%. This observation limits the application of such interfaces. The problem is attributed to the size of the holes and morphological changes at the insertion site due to long-term implantation.

To circumvent such difficulties, cone-ingrowth electrodes were developed. A gold wire is inserted in a hollow glass cone filled with nerve growth factor or another neurotrophic factor. The devices can be implanted for long term in the brain with remarkably stable activity. Furthermore, multielectrode contacts can be implanted in the motor areas of the brain in order to control movement or to develop real-time, electroencephalogram (EEG) based brain–computer interfaces, which function on processing the event-related synchronization or desynchronization of EEG wave patterns.

3.7.2 Microfluidic Structures

In the design of neuronal experiments, a promising microfluidic structure is the utilization of compartments with microscopic geometries for localized exposure of neurons to specific and controlled environmental conditions. Examination of certain neurodegenerative diseases such as multiple sclerosis (which causes injury to the axon) exemplifies the physiological importance of localization of neurons as opposed to whole neuron exposure, which is traditionally encountered in neuronal cultures. *In vitro* targeted studies of neurons are important in further understanding the molecular basis of neurodegenerative diseases.

Developing an optimized experimental platform is essential to conducting neuronal experiments. Scaling the experimental platform to match the cell geometry necessitates the use of miniaturized chemical stimulation and analysis systems. Achieving this analysis and simulation in a microsystem has long been an advertised promise of labs working with chips. As a result, there have been a number of developments in the application of microfluidic systems for neuronal experiments.[47,48]

Strategies for the use of microfluidic devices range from seeding neurons in reservoirs and growing axons into adjacent reservoirs (separated by microchannel barriers) to achieving localization of neurites and their isolation from the cell soma for targeted studies. Partially enclosed reservoirs separated by thin walls incorporating microchannels underneath the barriers have also been successful in providing isolation as well as improved access to the soma and axons.[49,50]

Molded chambered and channels have been experimentally proven to separate the soma with good adhesion between the poly(dimethylsiloxane) (PDMS) cover and the glass substrate with no leakage between the two reservoirs or between the channels (see Figure 3.12).

The development and spatial generation of the nervous system is highly regulated by the spatio-temporal regulation of guidance molecules. The

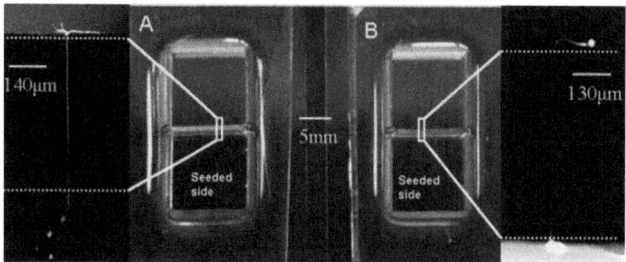

Figure 3.12 (A) Image of the planar glass device with channeled PDMS mold. (B) Planar PDMS with channeled glass—the zoom images show the cells growing through the channel into the adjacent compartment.[48] (Reprinted by kind of permission of the Royal Society of Chemistry.)

chemoaffinity of tip of the neurite plays an important role in directing the growth of neurites along specific pathways.[51] Neuronal wiring and guidance is a response to multiple attractive and repulsive cues operating at various spatial scales. Furthermore, there are elaborate time sensitive regulatory mechanisms which dictate whether a given cue elicits a response from the neurite or is ignored.[52] The concentration and gradient of growth factor ingredients are essential in providing guidance cues for neurite response.[53] Microfluidic devices are an important tool for studying neurite guidance under chemogradients as well as the tuning and promotion of neurite regeneration.[54]

The power of microfluidic devices to control the microenvironment around the neurons offers an experimental *in vitro* platform for studies that seek to investigate the multitude of growth cone cues. Microfluidic devices offer distinct advantages over other *in vitro* systems for studies investigating the effects of growth factors on neurite outgrowth.[55–57] Through use of microchannels and compartments, microfluidic devices offer user-defined control over the spatio-temporal distribution of growth factor in the extracellular matrix environment. Furthermore, by isolating and tracing individual neurites as they grow, these devices allow identification of cell-specific processes beyond the global response of the neuronal population to specific cues. These advantages can elucidate the complexities of physiologically relevant growth factor gradients and their specific cell signaling processes.

Aside from localized studies and compartmentalization of cells, use of microfluidic structures and valves have made cell co-culture platforms possible, where individual manipulation of the microenvironment of different cell types can be performed on the same experimental setup. The application of these microstructures enables new experimental processes to precisely control and manipulate the microenvironment of cells.[58] These co-culture platforms have been utilized to maintain healthy cultures of hippocampal neurons and glia for several weeks under optimal conditions.[59]

The microfluidic structures used in co-culture experiments rely on chambers with interconnected channels with dynamically controlled connections. An example of such structures is shown in Figure 3.13 where a structure is based on

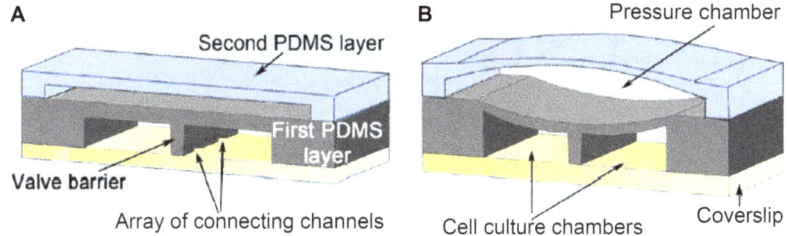

Figure 3.13 Schematic representation of valve barriers for reversible separation of microfluidic chambers A) pressure reduction opens the valve opening a microfluidic connection channel between the cell culture chambers B) Pressure build-up by filling the pressure chamber closes the valves, separating the cell culture compartments.[59]
(Reprinted by kind permission of Elsevier BV.)

fabricated soft lithography techniques using replica molding comprised of two layers of PDMS assembled on a glass cover slide. The valve operation to control the chamber connections operates by filling control channels with either water or air to create pneumatic or hydraulic pressure to provide a controllable barrier valve, thereby opening or closing the microchannels that connect to the chambers housing the cells.

Microfluidic platforms can be designed to support the culture of distinct cell populations in close proximity. Figure 3.14 shows an example of co-culturing of glial cells and neurons in microfluidic platforms. Using these platforms, glial cells can be cultured in isolation from neurons until reaching near confluence. Subsequently neuronal cells can be loaded into the other chamber and attached while the barrier valve remains closed. By adding fresh neuronal media to the chambers and opening the interconnection to allow flow of media from one chamber to the other, glia-conditioned media can be provided to the neurons.

3.7.3 Self-assembled Networks

Research has shown it is possible to generate networks composed of controllable low density neuronal cultures linked by bundles of axons and dendrites with predesigned geometry and topology. Through manufacturing of isolated islands of cell clusters surrounded by a non-adhesive background, stable self-wired engineered networks of both cortical and hippocampal neurons have been produced.[8] Self-assembled networks often rely on the natural tendency of cells to form efficient wiring between the closest neighboring islands rather than on predesigned direction. Cell positioning within the networks can be achieved through controlled network geometry, which can be designed with wiring between cells consisting of axons and dendrites.

A template for the network in self-assembled cultures can be lithographically defined, and the connectivity of the network is determined by the position of the cultured clusters and islands of cells. Cells on these platforms, illustrated in

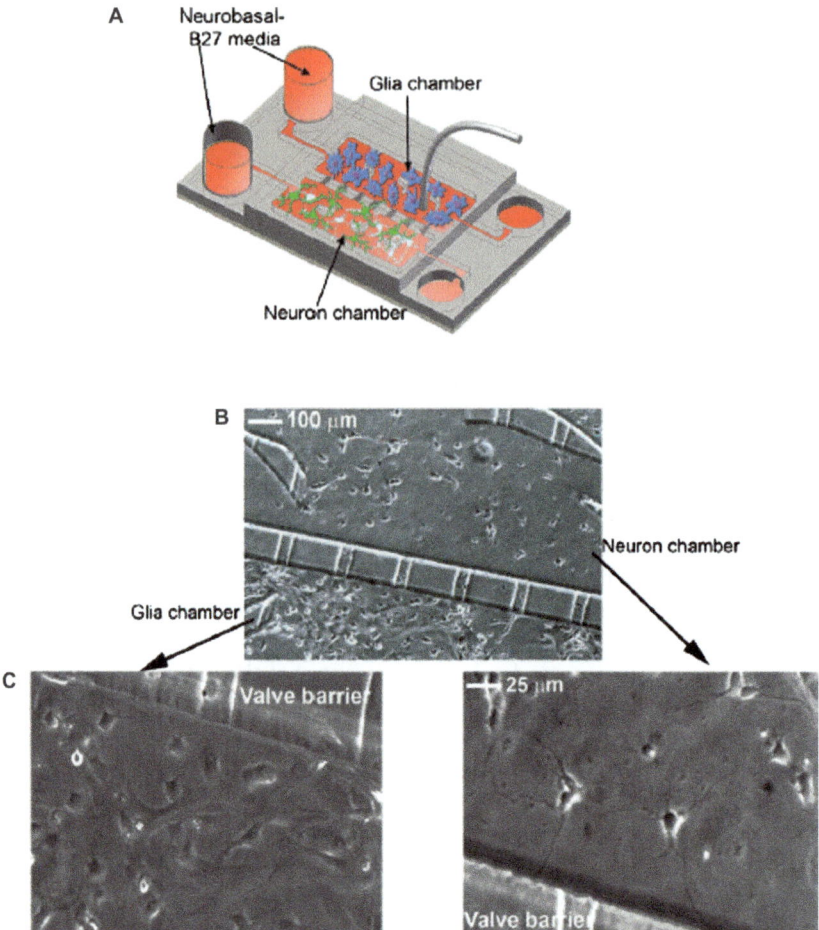

Figure 3.14 (A) Schematic representation of co-culturing of neurons and glial cells on a microfluidic platform with flow of from the glial side to the neuronal chamber established through higher culture media level on the glia side. (B) Image of co-culture platform with (C) showing a zoom image of the glia compartment on the left and neurons compartment on the right separated by the barrier valves.[59]
(Reprinted by kind permission of Elsevier BV.)

Figure 3.15, migrate on a low affinity substrate toward a high affinity substrate on adhesive templates, which they adhere to and assemble on.

3.8 *In Vitro* Microelectrodes Arrays

Specific measures are followed in order to study the activity and development of neuronal cell networks in culture or in brain slices. To begin with, planar biosensor arrays are built over embedded electrodes to overcome the challenges of probing a large number of neurons simultaneously using classic glass

Electronic Detection Techniques 111

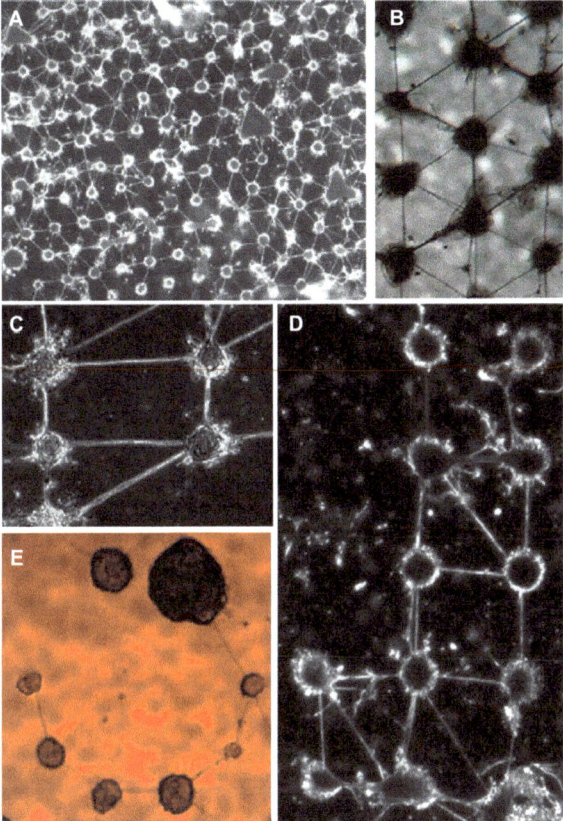

Figure 3.15 (A) Engineered networks of different cell geometries with PDL (poly-D-lysine) templates on glass (B–E). Low magnification image of illustrating interconnected ordered cell clusters designed lattices with (B) triangular lattice, (C) a single and row lattice, (D) square lattices and (E) circular arrangement. Adhesive sites were 100 μm in diameter and 'neuronal wiring' was achieved almost exclusively between neighboring islands.[8] (Reprinted by kind permission of the Institute of Physics Publishing.)

micropipettes. The smaller the electrodes, the higher the impedance, and so platinum-black electroplating lowers this electrode impedance to acceptable value levels of mega ohms. To improve the electrical resistance between neurons and electrodes, a cellular matrix (that promotes adhesion) is required.

Even so, some neurons will grow at a large distance from the recording electrodes and measurements will not feasible. To overcome this particular problem, electrode wells (to trap the somata of the cells) can be designed so that the neurons form a network by extending neuritis and connecting outside the wells, while the cell body remains in contact with the electrode at the bottom of the well. In order to guide the growing of the neurons on or around specific electrodes, some cells can be removed using a precise laser or employing neurophilic and neurophobic patterning. The patterning can be obtained by chemically depositing permissive hydrophilic layers (organosilanes, laminin, poly-D-lysine, polyethyleneimine) and non-permissive hydrophobic layers (polyimide, glass, fluorocarbon) into the desired pattern by lithographic, deposition or microstamping methods. The final goal is to design circuit architecture of neural networks with real neurons in the nodes to study the way real neurons in the network respond to input signals.

However, the reality is the connections made by the neurons are difficult to control and the function of the network is likely to be compromised. In addition to blocked ionic channels in the membrane (due to the way the cell sits on the surface or inside the well), distorted signals may also be generated both intracellular and extracellular. One possibility to overcome this challenge is to replace the neurotrophic factors in the implanted electrodes with living neuron cultures, which will establish dendritic connections to the brain tissue after implantation, fixing the probe in place and improving the interface with the nearby neurons.

The most interesting application of *in vivo* neural networks is the study of the cell metabolism and the reaction of the cells when exposed to different small molecules (*e.g.* pharmacological drugs, neurotrophic factors, neurotoxins) and other biochemical stimulants of relevant interest (GABA, nerve growth factor, *etc.*). In this way the spontaneous oscillatory activity of the living network and entrainment processes can be studied in greater detail to understand how a culture of neurons outside the brain organizes itself in such a coherent manner in the absence of a higher order control signal like the ones supposedly necessary to coordinate the brain activity in totality.

It has been shown that neurons are extremely sensitive to toxins and that the cellular death can be witnessed in cultures real time (including the degree of recovery after the toxin is removed from the system). Furthermore, it is hoped that these studies will one day render experimental testing with live animals unnecessary, which would be a remarkable accomplishment.

More importantly, the living neural network has been shown to be trainable and capable of learning in the same way that an artificial neural network can be trained to perform patterns of recognition. Such exciting applications are hoped to one day reveal the brain's cellular mechanisms for learning and cognitive abilities.

3.9 Microfluidics in Neurobiological Research

Microfluidic devices are emerging as powerful tools in neurobiology because of their ability to precisely control the spatial and temporal growth of neurons and manipulate the microenvironment of cells *in vitro*. Since their introduction in 1998, microfluidics device have been used intensively for cell biology studies. Miniaturization, integration and automation represent important potential advantages for exploring neuron cell attachments to substrates, neurite growth, nerve cell interactions, neural stem cell differentiation, neuropharmacology, neuroelectrophysiology, and biosensing applications. Although we present only a few selected applications below, growing, directing and interacting with cells on chips has been one of the flourishing areas in the field of microfluidics with many applications and several reviews.[60,61]

Microfluidic platforms allow the control and manipulation of the microenvironment of different cell types using low reagent and sample volumes, and achieving a level of specificity otherwise unachievable through traditional culturing techniques. Advances in microfabrication methods, such as integration of valves, pumps, mixing modules and controlled flow channels allow separation of the cell culture, control of culture media and fluidic exchange, as well as adjustable mechanical design of barriers between the chambers[59] and directed growth patterns of neurons.[62] The latest advances in surface chemistry, electrochemistry and microtechnology can be combined to create smart, low-cost and disposable devices, with spectacular applications in analyzing the structure and function of single neurons or networks of neurons to understand their activity and to elucidate mechanisms of memory and learning that may lead to the understanding of physiological and pathological brain processes.

Microfluidic platforms can be used to culture cells in separate compartments to conduct localized drug treatments. Such microcompartments can isolate axons for studying axon degeneration/regeneration under multiple parallel experimental conditions.[63] Figure 3.16 composed of three PDMS layers (a substrate layer, a compartment layer, and a culture media reservoir layer) is an example of such microcompartments for neuron cultures.

Prior to compartmetalization and culturing, microfluidic devices can also be used to isolate specific populations of cells derived from the nervous system. Examples include high throughput and high purity neural cell sorting through two-step separation by combining soft inertial microfluidics and pinched flow fractionation.[64] The concept of separation is illustrated in Figure 3.17. In this application separation is achieved in two steps. The first involves soft inertial microfluidics where cells experience a very high velocity gradient produced by

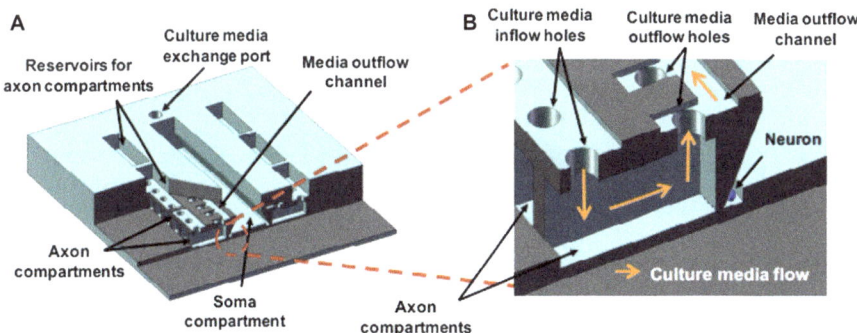

Figure 3.16 (A) Illustration of the assembled neuron culture platform device showing cross-sections. (B) Close-up view of the axon compartment showing culture media flow during the one-step culture media replenishing process.[63]
(Reprinted by kind permission of the Royal Society of Chemistry.)

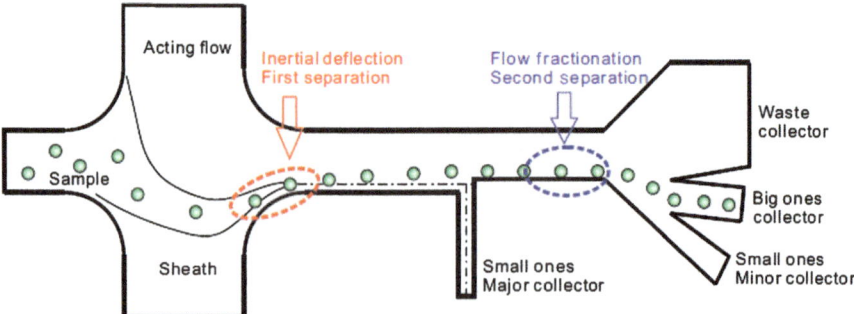

Figure 3.17 Two-step separation of neural cells.[64]
(Reprinted by kind permission of the Royal Society of Chemistry.)

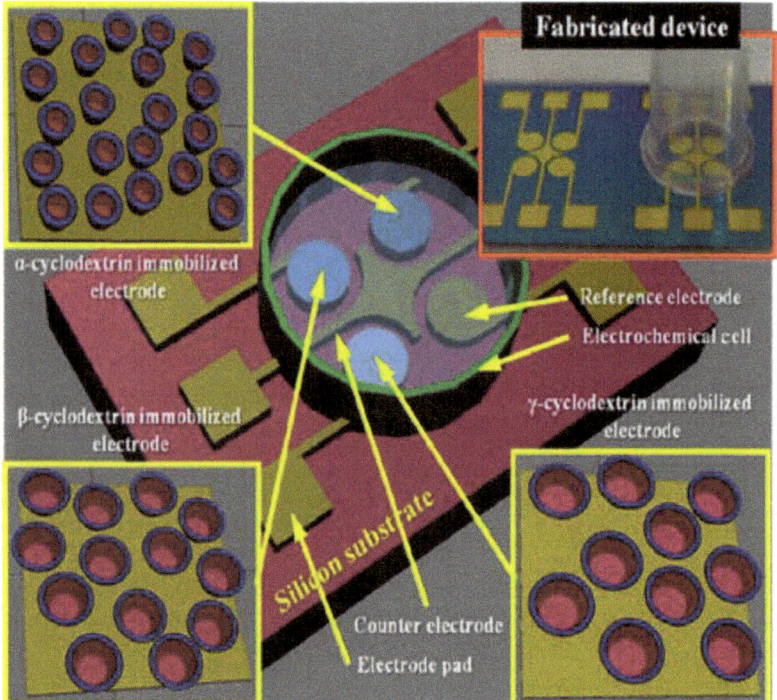

Figure 3.18 Illustration of modified electrodes array on silicon substrate for detecting catecholamine neurotransmitter.[65]
(Reprinted by kind permission of the Royal Society of Chemistry.)

three flow streams (sheath flow, sample flow and acting flow). Secondly at high gradient, the imbalanced velocity leads to a more rapid rotation of cells and subsequently a higher lifting force deflects larger particles from the original carrier fluid.

electrophysiological behavior efficiently, which may include comparison of tissue properties at different locations;
III. To monitor changes of electrical activity over periods of time not accessible with individual conventional electrodes (*e.g.* glass capillary or tungsten electrodes) in *in vitro* experiments.

Furthermore, the exceptional stability of the recording situation when MEAs are used allows analyses that would otherwise not be feasible. As the MEA technology can be applied to any electrogenic tissue (*i.e.* central and peripheral neurons, heart cells and muscle cells), the MEA biosensor is an ideal *in vitro* system to monitor both acute and chronic effects of drugs and toxins and to perform functional studies under physiological or induced pathophysiological conditions that mimic *in vivo* damage. By recording the electrical response of various locations on a tissue, a spatial map of drug effects at different sites can be generated which provides important clues about a drug's specificity.

Commercially available microarrays have been used to investigate the electrophysiological behavior of neuron cell populations for basic neuroscience studies or for applied neuropharmacological investigations. Different experimental protocols have been based on the use of an NMDA agonist and a non-NMDA agonist (α-amino-3-hydroxy-5-methyl-4-isoxazole propionic acid, AMPA) to modulate neuronal excitatory response consisting of an early rapid component, mediated by AMPA receptors, and of a slower component, mediated by NMDA receptors.[69] All the experimental sessions in the study by Martinoia *et al.* started with a 20 min recording of the network spontaneous activity, in physiological solution as control condition, before chemicals were added to the bath solution. For each culture, the effects of drugs on the endogenous glutamate efflux were measured as the percentage variation of endogenous glutamate amount in the presence of drugs with respect to the corresponding control value in the absence of drugs (*i.e.* the glutamate amount in the basal period immediately preceding the drug exposure period). In order to investigate modulations in the network electrophysiological activity both at the spike and at the burst level, it was realized that such modulations appear to be drug-specific and dose-dependent. In the case of high dose applications, the network showed an irreversible depression, indicating a possible dissolution of the network sensitivity and stability. Such an *in vitro* model of neuronal networks coupled to MEA devices is a potential high-sensitive biosensing system that can become a powerful and useful technique for drug screening applications. Unlike spinal cord neurons that represent a robust model in terms of network dynamics, cortical neurons represent a more interesting model, considering that it is likely to be the more advanced, adaptive and sensitive system. The results by Martinoia *et al.* further showed a modulation in the bursting and in the spiking network activity due to specific agonists acting on the glutamatergic ionotropic receptors (*i.e.* NMDA and AMPA). The designed protocols and experimental system showed the possibility of investigating the intervention of NMDA and non-NMDA receptors in spontaneous activity present in cortical neuronal networks under the effect of 'low-dose' application of different pharmacological substances that act specifically on glutamatergic synapses. Moreover, determination of the glutamate released by cortical neuronal networks in combination with the recording of integrated activity allowed correlation between the drug effects on the activity patterns and on functioning of glutamatergic excitatory synapses to be established. In fact, activation of high affinity NMDA receptors (by low micromolar NMDA concentrations) evoked changes of firing and bursting rate highly correlated to the effects on glutamatergic transmission, as assessed by glutamate release.

Cultured neural networks respond to neurotransmitters, blockers and many pharmacological substances such that *in vivo* behavior of neurons can be mimicked *in vitro*. For example, bicuculline blocks the main inhibitory neurotransmitter GABA, causing an increased spiking frequency, as does the chloride channel antagonist strychnine. Such cultures are cultured for long-term monitoring of neuronal electrophysiological activity including development, neuronal plasticity and axonal regeneration, all of which may be affected by the introduction of pharmaceutical compounds.

The peripheral nervous system has also been the focus of several studies.[77] Repetitive patterns of neural activity in spinal cord are responsible for behavior such as locomotion; they were investigated using neuronal networks of cells from the vertebrate spinal cord, and coupling properties were elucidated by using pharmacological methods. Bursting network activity and intrinsic activity in these cultures were modulated by 5-hydroxytryptamine (5-HT, serotonin) and the cholinergic agonist muscarine. Rhythmic activity in spinal cord slices has been monitored over long durations using MEAs.[70]

Current neurotoxicity assessment studies for new compounds are performed through *in vivo* neurobehavioral and neurotoxicity testing. Such testing procedures do not always accurately predict responses in humans, are costly, and require the use of numerous laboratory animals, the number of which could be limited by the introduction of *in vitro* methods.

3.10.4 Microelectrode Arrays in Toxicology

MEA technology is perfectly suited for the high throughput screening of toxic compounds. The recordings from multiple, separate cell networks can be made simultaneously from neuron cells situated in multi-well chips or plates. Compound toxicity can also be assessed in such electrophysiological approaches that allow .the transition from animal-intensive, descriptive toxicity testing to *in vitro*, predictive screening. The requirement for a better understanding of the potential hazards of the tens of thousands of chemicals currently used, as well as the necessity to increase the number of chemicals characterized for potential toxicity, are primary driving forces behind this change. The need to reduce the time, cost and numbers of animals used in contemporary toxicity tests also contributes to the shifts in approaches to hazard assessment. Ethical issues will also disappear. The need to extrapolate from animal to human models will be avoided, especially if human stem cells from renewable sources could be used. The future relies heavily on the use of *in vitro* assays, cellular and alternative species models, and predictive computational methods that incorporate knowledge about toxicity pathways.

The current approaches to the characterization of neurotoxic chemical hazards rely heavily on dose–effect characterization in whole animal models, analyzing behavioral (*e.g.* motor activity, water maze, functional–observational battery), neurophysiological (*e.g.* evoked potentials, EEG recordings, *etc.*) and/or pathological or structural (morphometric) endpoints. Hundreds of animals have to be tested using time-consuming behavioral and structural assessments. Research now focuses on obtaining information regarding cellular, molecular and submolecular actions of individual chemicals rather than the identification of toxicity pathways common to many chemicals. Neurophysiological assessments for compounds such as carbon disulfide, toluene solvents, metals and pesticides have been explored.[71] The blocking of type A GABA ($GABA_A$) receptors by a wide variety of insecticides increases excitability at the cellular level and results in increased network activity due to disinhibition of neuronal firing. This toxicity pathway ultimately underlies the acute neurotoxicity observed and neurophysiological approaches are ideal for a rapid detection of these types of toxicity.

Substrate-integrated microelectrode arrays provide important information regarding the neuron network structure and function that is difficult or impossible to obtain using other electrophysiological techniques. Some of the manufacturers that currently provide hardware and software for MEA systems include: Alpha MED Sciences, Osaka, Japan (MED systems formally manufactured by Panasonic); Axion Biosystems, Atlanta, GA; Ayanda Biosystems/SAS Biologics, Lausanne, Switzerland/Claix, France; 3-Brain, Landquart, Switzerland; MultiChannel Systems, Reutlingen Germany; Plexon, Inc., Dallas TX; and Tucker-Davis Technologies, Alachua, FL.[71]

MEA systems offer a great deal of flexibility regarding biological tissue and experimental design. A wide variety of electrically excitable biological tissues may be placed on the MEA. This includes cardiac tissue, primary cultures of nervous system tissue from many different regions, *i.e.* mouse frontal cortex cells, tissue slices (*e.g.* hippocampal slice) or retinas. MEA chips with characteristics specialized to the tissue can be made in-house or purchased commercially. The electrodes may be planar or they may have a 3D arrangement that allows penetration of the electrodes through outer layers of damaged cells in tissue slices, thereby enhancing the cell–electrode coupling. The spatial pattern of electrodes can also be changed, for example, to place more electrodes in specific regions of a hippocampal slice or retina. Following use, biological tissue is removed from the MEA chips, and thus they can be reused multiple times and with special care are quite economical ($10–30 per use). Figure 3.19 shows the assembling of an MEA (A–D) and the liquid handling robot over the culture chambers (E).

Electrical activity can be recorded either immediately (slices, retina) or following a growth phase (1–3 weeks) in culture. In neuronal cultures, activity consists of extracellular potentials generated by action potentials in cell bodies, axons and dendrites of neurons that are within the receptive field of an electrode on the array. The recorded activity is spontaneous but can also be

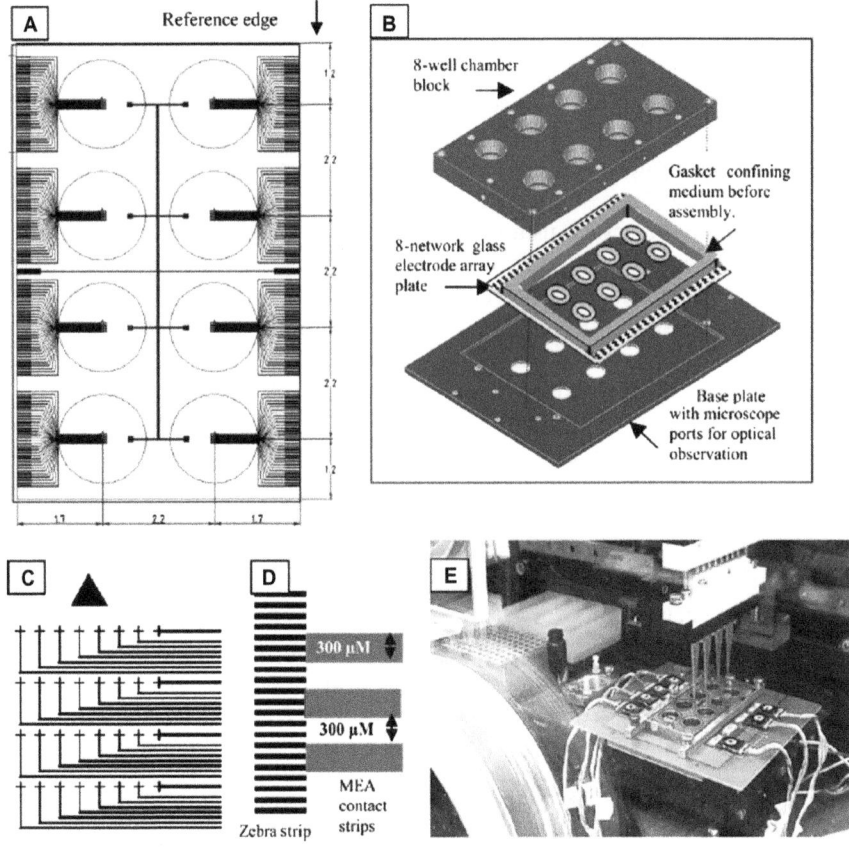

Figure 3.19 Assembly of recording MEA chambers (A–D) and the automatic liquid handling system (E).[71]
(Reprinted by kind permission of Elsevier BV.)

stimulated using MEA generators, where the MEA electrodes as stimulating electrodes. This allows the recording of evoked activity and makes possible the study of synaptic plasticity. The mechanisms of synaptic plasticity such as long-term potentiation and long-term depression have been well characterized in slices; this is not yet the case with cultures of dissociated neurons.

The events recorded by MEAs are extracellular potentials, but additional information can be extracted from the spatial and temporal aspects. The rate of action potentials ('spikes') and groups of action potentials ('bursts') are recorded on each electrode, as well as the overall network spike and burst rates. From this, many other parameters of spiking and bursting may be analyzed using custom written or commercially available software (*e.g.* Neuroexplorer from Nex Technologies, Littleton, MA), including burst duration, the number of spikes in a burst, the percentage of spikes in a burst, and the interspike and interburst intervals. Because spatial and temporal aspects of networks of neurons are examined, the analysis can also include more complex measures of network function such as autocorrelograms, cross-correlograms and sophisticated probabilistic models of network connectivity. Recently, an analysis of ignition sites (burst leaders) and leader-follower phase delay distributions has opened another window to the internal dynamics of networks and has revealed pharmacological sensitivities.[78] It has also been possible to determine dissociation constants for bicuculline, gabazine, and trimethylolpropane phosphate (TMPP) using frontal cortex cultures on MEAs with excellent agreement to data from more conventional assays.[79]

Table 3.1 forms the database of neuromodulators profiles 'finger prints' of concentration-dependent activity changes induced by a compound with known

Table 3.1 Database of substances whose 'fingerprint' activity has been established using MEA technologies.[71] (Reprinted by kind permission of Elsevier BV.)

Receptor agonists	*Anesthetics/analgesics*	*Gap junctions*
Acetylcholine	Fentanyl	Carbenoxolone
AMPA	Ketamine	1-Octanol
Baclofen	Levomethadone	Mefloquine
GABA	Morphine	Sodium propionate
Glycine	Phenobarbital	
NMDA	Propofol	*Phytopharmaceuticals*
Neuropeptide Y	Remifentanil	Passion flower
Norepinephrine	Sufentanil	St Johns Worth
Pregnanolone	Thiopental	*Valeriana*
Substance P	Tramadol	
		Particles and ions, metals
Receptor antagonists	*Anticonvalsants/sedatives*	TiO_2, C, Fe_2O_3, As_2O_3,
APV	Carbamazepine	As_2O_5, MeAs(III),
Atropine	Clonazepam	MeAs(V), Mg^{2+}, Mn^{2+},
Benzoquinone	Diazepam	La^{3+}, C_3H_9ClSn
Bicuculline	Flunitrazepam	
Flumazenil	Gabapentin	*Other*
L-512,804	Levetiracetam	Aβ42
L-733,060	Phenytoin	Albumin
MK 801	Topiramate	Ammonia
Muscimol	Valproate	Apomorphine
NBQX	Zolpidem	Cisplatinum
Picrotoxin		Cortisol
SCH50911	*Opioids*	Cyclosporine A
SCS	Dermorphine	DEET
Strychnine	DAMGO	DMSO
	DPDPE	EBAAP
Antidepressants/neuroleptics	Dynorphin A	Ethanol
Amitriptyline	Enkephaline	Lithium heparin
Apomorphine	Endomorphine I + II	Neurotensin
Clozapine	Nalorphine	Nocodazole
Fluoxetine	Naloxone	Tacrolimus
Fluvoxamine	Nociceptin	Tetrodotoxin
Haloperidol	U50488	Wortmannin
LiCl		

single or multiple modes of action covering most of the neurophysiologically relevant pathways. This allows the comparison of the response of unknown new substances to those of substances in the database. Substances with uncertain modes of action can be identified and compounds classified accordingly, providing information on the toxicity pathway through which these compounds may act. Therefore, analysis of spike train data using pattern recognition methods provides a rich description of a wide variety of parameters for the purpose of high-content screening.

Finally, in addition to analysis of spatio-temporal aspects of network activity, the individual waveforms of each action potential event can be collected and analyzed for changes in amplitude and/or duration. Thus, MEAs are high-content platforms that can provide detailed information regarding the changes in function of networks of neurons exposed to test compounds. The activity patterns in response to drugs or toxins are remarkably reproducible.

Future high-throughput, parallel and rapid physiological screening will require large-scale platforms, served by liquid handling robots connected to automated data acquisition and analysis systems. Such scaling can be achieved by multiplying the number of recording modules or by expanding the number of networks per module. Thousands of toxins and pharmacological substances can be tested in parallel in a very short time. No other technology has the potential to offer rapid collection of information on this scale.

3.10.5 Microelectrode Arrays in Basic Neuroscience Research

We are still overwhelmed by the complexity of the human brain and elucidating the disorders of the brain is a pressing issue due to the economic and social consequences of degenerative disorders. It is known that conditions such as Alzheimer's disease (AD) and epilepsy depend on the concerted activity of neural networks. Use of cultures as surrogates for *in vivo* experiments relies on the premise that the 'functional units' of the brain are distinct cellular populations; cultured networks of neurons can be considered the 'functional unit of functional organization' in the brain. MEA technologies can facilitate the study of brain disorders and pathologies.

3.10.5.1 *Alzheimer's disease*

Brain degeneration that leads to dementia is a heartbreaking process: patients do not recognize their own children and they lose their identity. There is an immense emotional and financial pressure on families and society. No objective clinical detecting tests are available (only behavioral and cognitive scores) and no cure for the disease. One form of dementia, Alzheimer's disease (AD) is a progressive neurodegenerative disease characterized by the cerebral accumulation of insoluble proteinaceous clusters of amyloid β-peptides (Aβ), which have been implicated in neuronal toxicity and deterioration. Consequently, efforts to mitigate neurodegeneration have emphasized the reduction of Aβ aggregates.

Four medications are currently approved by regulatory agencies such as the US Food and Drug Administration (FDA) and the European Medicines Agency (EMEA) to treat the cognitive manifestations of AD. Three are acetylcholinesterase inhibitors (Aricept, galantamine and rivastigmine) and the other is memantine, an NMDA receptor antagonist. No drug has an indication for delaying or halting the progression of the disease.

Aricept, galantamine and rivastigmine work by enhancing cholinergic function. This is accomplished by increasing the concentration of acetylcholine through reversible inhibition of its hydrolysis by acetylcholinesterase. Memantine acts on the glutamatergic system by blocking NMDA glutamate receptors and blocks the toxic effects associated with excess glutamate and regulates glutamate activation. New drugs still in trials such as Bapineuzumab, Solanezumab, Gammagard are antibodies against amyloid (the first two lab-made and the last one culled from blood) and extremely expensive.

Cytotoxic effects of Aβ interactions with neuron cells are expressed by the accumulation on ganglioside and cholesterol-rich domains of cell membranes demonstrating localized neurite degeneration prior to induced cell death, loss of Ca^{2+} homeostasis, which may result from changes in endogenous ion transport systems (e.g. Ca^{2+} and K^+ channels and Na^+/K^+-ATPase) and membrane receptor proteins, such as ligand-driven ion channels and G-protein driven releases of second messengers, or the formation of heterogeneous ion channels. These changes in membrane transport systems are proposed to impair neuronal function by compromising membrane integrity and increasing its ion permeability. Other cytotoxic effects are represented by the generation of reactive oxygen species leading to peroxidation of phospholipids in the membrane, inhibition of phosphorylation, and reduction of ATP levels and cytoplasmic pH, and reduced cellular reductase activity (a measure of cellular viability). The treatments are plagued by side effects (brain edema, cardiovascular, gastrointestinal, etc.), which sometimes determine the immediate cessation of treatment in order to spare the patient's life.

Soluble oligomers of Aβ are considered to be one of the major contributing factors to the development of AD. Most therapeutic development studies have focused on toxicity directly at the synapse. Soluble Aβ can also cause functional toxicity; namely it inhibits spontaneous firing of hippocampal neurons without significant cell death at low concentrations. This toxicity eventually leads to the loss of synapses as well, but the effect on function may precede this loss by a considerable amount of time.

One of the characteristic symptoms of AD is the impairment of memory function. Although memory mechanisms are still not well understood, two processes—long-term potentiation (LTP) and long-term depression (LTD)—are thought to be opposing forces that are responsible for memory storage. Aβ oligomers inhibit LTP, enhance LTD and reduce dendritic density in the normal hippocampus. The toxic effects of Aβ oligomers were demonstrated by using cultured hippocampal neurons on an MEA, confirming previous data from patch clamp studies. Using multielectrode arrays this process can be investigated in a very elegant way, as described by Varghese et al.[72]

Aβ oligomers were synthesized using an established protocol and characterized using western blots. Embryonic hippocampal neurons were plated on MEAs and after 14 days *in vitro* were treated with either 100 nM Aβ oligomers alone or a combination of Aβ and curcumin, an Aβ oligomerization inhibitor. Recordings were obtained over this period of time in order to determine the time course of both compounds. The observed results were compared with patch clamp data obtained from similar experiments. Cell death was quantified using a live–dead assay. When applied to hippocampal neurons cultured on MEAs, Aβ had a pronounced effect on the spontaneous firing of the cells, even at concentrations in the nanomolar range. Treatment with Aβ stopped spontaneous activity completely and the time until cessation was concentration dependent. Curcumin was able to partially reverse the loss of spontaneous activity, in accordance with other patch clamp experiments.[80]

This study demonstrated that it is possible to develop a high-throughput MEA screen for the measurement of a drug's effect on functional toxicity of low concentrations of Aβ and this model was the first step for an *in vitro* functional model of the development of AD.

3.10.5.2 Epilepsy

Epilepsy is a serious disorder that affects 50 million people globally, a third of whom are refractory to currently available treatments. Screening of potential anticonvulsants takes many forms and employs numerous *in vitro* and *in vivo* models of epileptiform activity and seizures, respectively, ranging from cultured neurons to whole animals. Additionally, extracellular and intracellular electrophysiological recordings from intact brain slices are important techniques for highlighting potential anti-epileptic properties. The removal of extracellular Mg^{2+} ions from bathing fluid or the addition of the potassium channel blocker 4-aminopyridine (4-AP) can induce spontaneous epileptiform activity in a variety of brain slice preparations. This bioelectrical activity is analogous to *in vivo* status epilepticus as measured by EEG.

Hill *et al.*[73] have developed multielectrode arrays recording the Mg^{2+}-free and 4-AP models of epileptiform activity in neurophysiologically mature young adult rat brain slices. MEAs allow extracellular recording from many electrodes across a large area of the slice, assessing parameters including burst frequency, duration and amplitude. MEAs also allow these parameters to be measured at multiple points, better highlighting site-specific differences and patterns of neuronal activity across the slice. Furthermore, important information on burst origin, propagation paths, speeds and focal points can be easily and robustly obtained using MEAs. These are important properties of epileptiform activity that are not easily or accurately assessed by other means. The method was validated by examining the effects of two anticonvulsants (felbamate and phenobarbital) that have been previously reported to exert significant effects in models using conventional extracellular recording. Burst frequency and amplitude changed over time in both models in the absence of anticonvulsant drugs. This is a finding that has significant implications for conventional, as well as MEA, recordings of epileptiform activity in slices. Felbamate and phenobarbital effects upon the overall power of the recorded signal were assessed using power spectral density (PSD) measures. The speed and paths of propagation were also assessed and compared in the absence and presence of phenobarbital and felbamate. Hill *et al.* have demonstrated validation of MEAs method in comparison with previous extracellular recording data from the same models, and also described previously unreported background changes in burst characteristics important to the use of these models. MEA recording of epileptiform slice activity increases the depth and quality of data gathered, and represents an important method for the early *in vitro* screening of novel antiepileptic drugs.

3.10.5.3 Microelectrode Arrays for Retinal Studies

One interesting application is related to the exploration of explanted retina tissue using MEAs.[74] The retina is a peripheral, easily accessible part of the central nervous system. Stimulation by light results in a complex signaling by neurons within the layers of the retina. The retinal ganglion cells transmit retinal information to higher visual centers in the brain *via* their axons that form the optic nerve. Retinal function can be affected by acute injuries, intoxications or retinal diseases, either inherited or acquired, resulting in visual impairment or even blindness.

In clinical practice the so-called electroretinogram (ERG) is a widely used ocular electrophysiological test to diagnose impaired vision. Light impulses falling on the retina synchronously activate a large number of neurons and Muller glia cells, which regulate the extracellular potassium concentration. The resulting change in trans-retinal voltage is measured as the ERG. The ERG has a multiphasic waveform. Its shape depends mainly on the stimulus conditions, the state of the retina's adaptation and the species. The full-field ERG of a dark adapted retina in response to a bright flash of white light consists of four major components—the fundamental a-, b- and c-wave at light onset and the d-wave at light offset. Each of the components can be attributed to the activity of certain retinal cells. Under pathophysiological conditions the shape and amplitude of these components is altered and can be influenced by pharmacological compounds.

A retina sensor, based on multisite recording of local ERGs *in vitro*, has been developed to easily and effectively assess effects of pharmacological compounds and putative therapeutic, drug side effects, and the consequences of degeneration-related processes on retinal signaling (Figure 3.20).

For the recording of light-evoked activity, a retinal segment with the pigment epithelium, dissected from an explanted chicken retina, is placed ganglion cell site down on a MEA. Local ERGs (microERGs) with the typical components and ganglion cell spikes can be recorded with the appropriate filter settings. The prominent components of the microERG can be pharmacologicaly identified for the b-wave, which is smaller in recordings from isolated retinas than in recordings from intact eyes. During superfusion with drugs in defined concentrations, specific alteration of the ERGs can be monitored. In its present

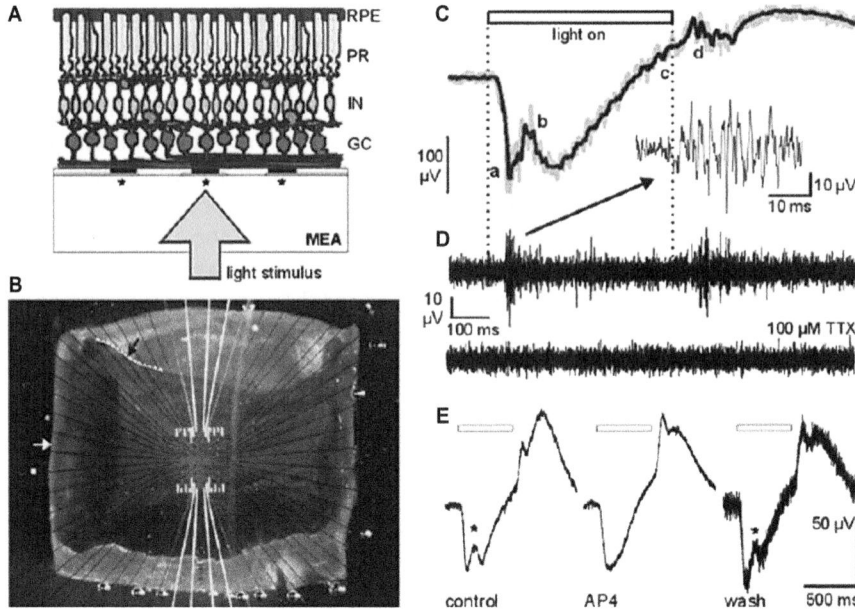

Figure 3.20 MEA retinasensor. (A) An explanted retina is placed on an MEA with the ganglion cell side (RPE, retinal pigment epithelium; PR, photoreceptor; IN, interneurons; GC, ganglion cells). The light stimulus is projected through the transparent MEA and retina onto the photoreceptors. Ganglion cell activity and retinal field potentials (microERG) are recorded by the substrate-integrated microelectrodes (*). (B) View through an MEA on a retinal segment (white arrow) from a chicken retina. The broken line marks the border of the pigment epithelium. (C) microERG with a-, b-, c- and d-wave of a chicken retina, evoked by full-field white light, impulse duration 500 ms, 0.5 Hz. Grey curve denotes a single sweep, 0.5 Hz to 2.8 kHz, the black line five sweeps averaged, filtered 0.5–100 Hz. (D) Spike activity (insert), extracted from the grey curve shown in (C) by off-line filtering at 200 Hz to 2.8 kHz. The spikes were extinguished by 100 µM TTX. (E) Drug action on a light-evoked microERG. 2-Aminophosphonobutyric acid (AP4), a blocker of the on-signal pathway in the retina, resulted in the disappearance of the b-wave (*) that mainly reflects retinal Muller cell and bipolar cell activity. B-wave amplitude is restored after washing out the drug. Light pulses 500 ms, 0.5 Hz.[74] (Reprinted by kind permission of Springer.)

form, the MEA retina sensor is suitable for drug testing over several hours, depending on the reversibility of tested drug effects, but long-term monitoring of drug effects is also possible.

References

1. P. L. Nunez and R. Srinivasan, *Electric Fields of the Brain: The Neurophysics of EEG*, 2nd ed., Oxford University Press, Oxford, 2006.
2. G. R. Holt and C. Koch, *J. Comput. Neurosci.*, 1999, **6**, 169.

3. E. Claverol-Tinture and J. Pine, *J. Neurosci. Methods*, 2002, **117**, 13–21.
4. L. Xiangning, Z. Wei, L. Man, Z. Shaoqum and L. Qingming, *Prog. Nat. Sci.*, 2006, **16**, 1337.
5. S. Marom and G. Shahaf, *Q. Rev. Biophys.*, 2002, **35**, 63.
6. J. V. Selinger, J. J. Pancrazio and G. W. Gross, *Biosens. Bioelectron.*, 2004, **19**, 675.
7. O. Shefi, I. Golding, R. Segev, E. Ben-Jacob and A. Ayali, *Phys. Rev. E: Stat. Nonlin. Soft Matter Phys.*, 2002, **66**, 021905.
8. R. Sorkin, T. Gabay, P. Blinder, D. Baranes, E. Ben-Jacob and Y. Hanein, *J Neural Eng.*, 2006, **3**, 95.
9. P. Lamoureux, J. Zheng, R. E. Buxbaum and S. R. Heidemann, *J. Cell. Biol.*, 1992, **118**, 655.
10. R. G. Thakar, F. Ho, N. F. Huang, D. Liepmann and S. Li, *Biochem. Biophys. Res. Commun.*, 2003, **307**, 883.
11. D. Falconnet, G. Csucs, H. Michelle Grandin and M. Textor, *Biomaterials*, 2006, **27**, 3044.
12. R. Kriparamanan, P. Aswath, A. Zhou, L. Tang and K. T. Nguyen, *J. Nanosci. Nanotechnol.*, 2006, **6**, 1905.
13. F. L. Yap and Y. Zhang, *Biosens. Bioelectron.*, 2007, **22**, 775.
14. H. Shi, W.-B. Tsai, M. D. Garrison, S. Ferrari and B. D. Ratner, *Nature*, 1999, **398**, 593.
15. S. Khan and G. Newaz, *J. Biomed. Mater. Res. A*, 2010, **93A**, 1209.
16. M. Singer, R. H. Nordlander and M. Egar, *J. Comp. Neurol.*, 1979, **185**, 1.
17. C. R. Norris and K. Kalil, *J. Neurosci.*, 1991, **11**, 3481.
18. S. Kaech and G. Banker, *Nat. Protoc.*, 2006, **1**, 2406.
19. W. L. C. Rutten, *Annu. Rev. Biomed. Eng.*, 2002, **4**, 407.
20. A. Cohen, J. Shappir, S. Yitzchaik and M. E. Spira, *Biosens. Bioelectron.*, 2006, **22**, 656.
21. W. L. C. Rutten, H. J. van Wier and J. H. M. Put, *IEEE Trans. Biomed. Eng.*, 1991, **38**, 192.
22. C. Q. Huang, R. K. Shepherd, P. M. Center, P. M. Seligman and B. Tabor, *IEEE Trans. Biomed. Eng.*, 1999, **46**, 461.
23. P. Fromherz, *ChemPhysChem*, 2002, **3**, 276.
24. F. Heer, S. Hafizovic, T. Ugniwenko, U. Frey, W. Franks, E. Perriard, J.-C. Perriard, A. Blau, C. Ziegler and A. Hierlemann, *Biosens. Bioelectron.*, 2006, **22**, 2546.
25. F. Patolsky, B. P. Timko, G. Yu, Y. Fang, A. B. Greytak, G. Zheng and C. M. Lieber, *Science*, 2006, **313**, 1100.
26. I. Willner and E. Katz, *Bioelectronics*, John Wiley & Sons, Chichester, 2006.
27. J. L. Walker, *Anal. Chem.*, 1971, **43**, 89A.
28. C. A. Thomas Jr., P. A. Springer, G. E. Loeb, Y. Berwald-Netter and L. M. Okun, *Exp. Cell Res.*, 1972, **74**, 61.
29. G. W. Gross, *IEEE Trans. Biomed. Eng.*, 1979, **26**, 273.
30. J. Pine, *J. Neurosci. Methods*, 1980, **2**, 19.

31. U. Frey, U. Egert, F. Heer, S. Hafizovic and A. Hierlemann, *Biosens. Bioelectron.*, 2009, **24**, 2191.
32. W. L. C. Rutten, T. A. Frieswijk, J. P. A. Smit, T. H. Rozijn and J. H. Meier, *Biosens. Bioelectron.*, 1995, **10**, 141.
33. T. A. Frieswijk, W. L. C. Rutten, H. G. Kerkhoff and K. Lippe, in *Engineering in Medicine and Biology Society, Proceedings IEEE 17th Annual Conference 1995*, 1995, vol. 2, pp. 1103–1104.
34. L. Berdondini, M. Chiappalone, P. D. van der Wal, K. Imfeld, N. F. de Rooij, M. Koudelka-Hep, M. Tedesco, S. Martinoia, J. van Pelt, G. Le Masson and A. Garenne, *Sens.Actuators, B*, 2006, **114**, 530.
35. T. DeMarse, A. Cadotte, P. Douglas, P. He and V. Trinh, in *Proceedings 26th Annual International Conference of the IEEE Engineering in Medicine and Biology Society (IEMBS '04)*, 2004, vol. 2, pp. 5340–5343.
36. G. W. Gross, A. Harsch, B. K. Rhoades and W. Göpel, *Biosens. Bioelectron.*, 1997, **12**, 373.
37. A. Scarlatos, A. J. Cadotte, T. B. DeMarse and B. A. Welt, *J. Food Sci.*, 2008, **73**, E129.
38. S. Illes, W. Fleischer, M. Siebler, H.-P. Hartung and M. Dihné, *Exp. Neurol.*, 2007, **207**, 171.
39. P. Fromherz and M. Voelker, *Science*, 2009, **323**, 1429.
40. M. Voelker and P. Fromherz, *Small*, 2005, **1**, 206.
41. F. O. Morin, Y. Takamura and E. Tamiya, *J. Biosci. Bioeng.*, 2005, **100**, 131.
42. J. P. A. Smit, W. L. C. Rutten and H. B. K. Boom, *IEEE Trans. Rehabil. Eng.*, 1999, **7**, 399.
43. A. Branner, R. B. Stein and R. A. Normann, *J. Neurophysiol.*, 2001, **85**, 1585.
44. S. M. Potter, *Prog. Brain Res.*, 2001, **130**, 49.
45. A. Blau, C. Ziegler, M. Heyer, F. Endres, G. Schwitzgebel, T. Matthies, T. Stieglitz, J.-U. Meyer and W. Göpel, *Biosens. Bioelectron.*, 1997, **12**, 883.
46. Q. Zhao, J. Drott, T. Laurell, L. Wallman, K. Lindström, L. M. Bjursten, G. Lundborg, L. Montelius and N. Danielsen, *Biomaterials*, 1997, **18**, 75.
47. A. Manz, N. Graber and H. M. Widmer, *Sens. Actuators, B*, 1990, **1**, 244.
48. M. Arundell, V. H. Perry and T. A. Newman, *Lab Chip*, 2011, **11**, 3001.
49. I. Yang, R. Siddique, S. Hosmane, N. Thakor and A. Höke, *Exp. Neurol.*, 2009, **218**, 124.
50. S. K. Ravula, M. S. Wang, S. A. Asress, J. D. Glass and A. Bruno Frazier, *J. Neurosci. Methods*, 2007, **159**, 78.
51. B. J. Dickson, *Science*, 2002, **298**, 1959.
52. M. Tessier-Lavigne and C. S. Goodman, *Science*, 1996, **274**, 1123.
53. G. Curinga and G. M. Smith, *Exp. Neurol.*, 2008, **209**, 333.
54. C. R. Kothapalli, E. van Veen, S. de Valence, S. Chung, I. K. Zervantonakis, F. B. Gertler and R. D. Kamm, *Lab Chip*, 2011, **11**, 497.
55. H. Chen, Z. He, A. Bagri and M. Tessier-Lavigne, *Neuron*, 1998, **21**, 1283.
56. W. J. Rosoff, J. S. Urbach, M. A. Esrick, R. G. McAllister, L. J. Richards and G. J. Goodhill, *Nat. Neurosci.*, 2004, **7**, 678.

57. B. Genç, P. H. Özdinler, A. E. Mendoza and R. S. Erzurumlu, *PLoS Biol*, 2004, **2**, e403.
58. A. M. Taylor, M. Blurton-Jones, S. W. Rhee, D. H. Cribbs, C. W. Cotman and N. L. Jeon, *Nat. Methods*, 2005, **2**, 599.
59. D. Majumdar, Y. Gao, D. Li and D. J. Webb, *J. Neurosci. Methods*, 2011, **196**, 38.
60. J. El-Ali, P. K. Sorger and K. F. Jensen, *Nature*, 2006, **442**, 403.
61. A. M. Taylor and N. L. Jeon, *Crit. Rev. Biomed. Eng.*, 2011, **39**, 185.
62. R. Dong, R. P. Molloy, M. Lindau and C. K. Ober, *Biomacromolecules*, 2010, **11**, 2027.
63. J. Park, H. Koito, J. Li and A. Han, in *Proceeedings 14th International Conference on Miniaturized Systems for Chemistry and Life Sciences (µTAS 2010)*, Groningen, The Netherlands, 2010, pp. 860–862.
64. Z. Wu, G. Wicher, Å. F. Svenningsen and K. Hjort, in *Proceeedings 14th International Conference on Miniaturized Systems for Chemistry and Life Sciences (µTAS 2010)*, Groningen, The Netherlands, 2010, pp. 917–919.
65. J.-H. Yang, J. W. Park and H. Kim, in *Proceeedings 14th International Conference on Miniaturized Systems for Chemistry and Life Sciences (µTAS 2010)*, Groningen, The Netherlands, 2010, pp. 599–601.
66. A. M. Aravanis, B. D. DeBusschere, A. J. Chruscinski, K. H. Gilchrist, B. K. Kobilka and G. T. Kovacs, *Biosens. Bioelectron.*, 2001, **16**, 571.
67. A. Ghanem and M. L. Shuler, *Biotechnol. Prog.*, 2000, **16**, 471.
68. N. A. Mufti and M. L. Shuler, *Toxicol. In Vitro*, 1998, **12**, 259.
69. S. Martinoia, L. Bonzano, M. Chiappalone, M. Tedesco, M. Marcoli and G. Maura, *Biosens. Bioelectron.*, 2005, **20**, 2071.
70. A. Tscherter, M. O. Heuschkel, P. Renaud and J. Streit, *Eur. J. Neurosci.*, 2001, **14**, 179.
71. A. F. M. Johnstone, G. W. Gross, D. G. Weiss, O. H.-U. Schroeder, A. Gramowski and T. J. Shafer, *Neurotoxicology*, 2010, **31**, 331.
72. K. Varghese, P. Molnar, M. Das, M. S. Kindy, B. C. Wheeler and J. J. Hickman, *Alzheimers Dement.*, 2009, **5**, P428.
73. A. J. Hill, N. A. Jones, C. M. Williams, G. J. Stephens and B. J. Whalley, *J. Neurosci. Methods*, 2010, **185**, 246.
74. A. Stett, U. Egert, E. Guenther, F. Hofmann, T. Meyer, W. Nisch and H. Haemmerle, *Anal. Bioanal. Chem.*, 2003, **377**, 486.
75. G. W. Gross, A. Harsch, B. K. Rhoades, and W. Göpel, *Biosens. Bioelectron.*, 1997, **12**(5), 373–393.
76. T. H. Park and M. L. Shuler, *Biotechnol. Prog.*, 2003, **19**(2), 243–253.
77. X. Navarro, T. B. Krueger, N. Lago, S. Micera, T. Stieglitz, and P. Dario, *J. Peripher. Nerv. Syst.*, 2005, **10**(3), 229–258.
78. M. I. Ham, L. M. Bettencourt, F. D. McDaniel, and G. W. Gross, *J. Comput. Neurosci.*, 2008, **24**(3), 346–357.
79. S. O. Rijal and G. W. Gross, *J. Neurosci. Methods*, 2008, **173**(2), 183–192.
80. K. Varghese, P. Molnar, M. Das, N. Bhargava, S. Lambert, M. S. Kindy, and J. J. Hickman, *Plos One*, 2010, **5**(1).

CHAPTER 4
Nanosensing the Brain

Understanding how the brain works is one of the greatest scientific challenges of modern science. There is no general theory of brain function. The state-of-the-art plan in neurobiology is to find neurons or subcircuits of interconnected neurons and the neural correlation of the rate and manner of firing with behaviour and conscious experiences. Neuroscientists are still dreaming of finding the neural code to discover if mean firing rates are associated with experiences and to understand how the chemistry of neurotransmitters influences behaviour. While the function of the brain and consciousness itself are considered to be emerging from the complex, multi-hierarchical structure of the brain, it seems to also be an intrinsic property of the brain. The brain is self-wiring and then prunes the unwanted connections in the first years of life. The same phenomenon can be observed directly *in vitro* in culture of neurons using charge-coupled device (CCD) cameras. After establishing connections, some of them are pruned, according to unknown rules. Inside the brain, the whole process is a continuous and dynamic manifestation of brain's neuroplasticity, and the connectivity matrix is perpetually rearranging as individual neurons are making synaptic contacts and receiving synaptic contacts from thousands of other neurons. Only after understanding how this extremely complicated process unfolds can an attempt be made to restore the pattern of normal activity and to remove the pathological manifestations of the injured or diseased brain.

The latest discoveries using two-photon excitation techniques, fluorescent imaging including voltage-based functional imaging, functional magnetic resonance imaging (fMRI) and magnetoencephalography (MEG) capture images of brain activity without having the spatial or temporal resolution to detect the detailed patterns of neuronal firings. As such, new sensors with better time resolution, better signal-to-noise ratio, less photodamage and less invasive techniques are necessary. Newly discovered nanomaterials and the miniaturization of electronic circuits to the present 22 nm (Intel) and progressing to 14 nm and 10 nm (as Moore's law predicted, a size reduction of

RSC Detection Science Series No. 1
Sensor Technology in Neuroscience
By Michael Thompson, Larisa-Emilia Cheran and Saman Sadeghi
© M Thompson, L-E Cheran and S Sadeghi 2013
Published by the Royal Society of Chemistry, www.rsc.org

1.5 fold per year)[1] will open unimagined possibilities in neuroscience research. Neuroscientists must become aware of these exciting new techniques and use them in interdisciplinary teams where the gaps between electronic technology and neurology can be successfully bridged.

4.1 Nanoparticles as Reporters of Brain Activity

A new area of advanced detection methods in neurobiology is the development of nanoparticles. As introduced briefly in Chapter 1, nanoparticles are represented by small (5–8 nm) inorganic compounds made of semiconductor or metallic materials with well-defined quantum states and electronic structure. For example, they are composed of a metal core (cadmium, selenium, cadmium telluride), a zinc sulfate shell and an outer coating functionalized using bioactive molecules (Figure 4.1). Fluorescent quantum dots are used to visualize molecular processes in neuron cells using fluorescent microscopy methods. Their properties can be adjusted by changing the composition or size. Small changes in the radius translate into distinct color changes so they can be used to replace bulky organic fluorophores which interfere with the molecular structure of the object of investigation. The design of these nanostructures is based on the ability to control plasmonic behavior in metallic nanoparticles, quantum size effects in semiconductor heterostructures with designed asymmetries, and nanoparticles with implanted dopants possessing sharp emission spectra. Their small sizes allow large and specific energy jumps between the energy band gaps of excited electrons or electron-hole pairs.[2]

These inorganic nanoparticle optical probes can be tuned to match the photon energy requirements of the various excitation and detection systems. Unlike organic optical probes, they are photochemically robust during extended interrogation. For neuroscience studies, nanoparticles are combined

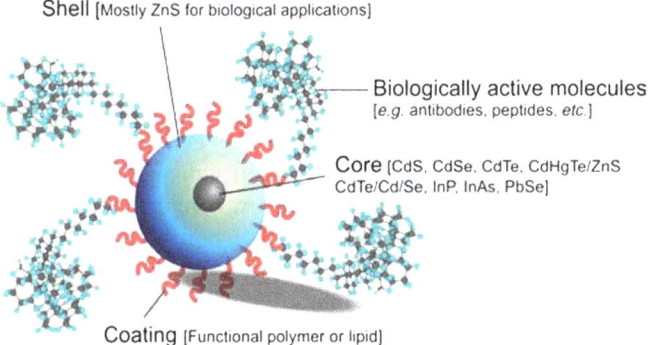

Figure 4.1 Structure of a semiconductor fluorescent quantum dot nanocrystal. The heavy metal core is responsible for the fluorescence properties of the quantum dot. The non-emissive shell stabilizes the core, whereas the coating layer provides anchor sites to organic and biological ligands such as antibodies, peptides and other organic molecules.[2]
(Reprinted by kind permission of the Society for Neuroscience.)

with organic nanostructures that are designed specifically for biofunctionalization, with the overall goal of attaching them within neural cells configurations. The structure of semiconducting nanoparticles enables the generation of excitons, which are very sensitive to the external electric field. This sensitivity can turn these nanoparticles into reporters with externally modulated fluorescence intensity spectra. They may be combined with selective molecular binding moieties to confer sensitivity to changes in local neurotransmitter concentrations. Quantum dots can be used as local optical reporters for neuroscience, and for visualizing dynamic molecular processes in neurons and glia on a large time scale, starting from seconds to many minutes, and on the small size scale of the synaptic cleft (20 nm) of neuron–neuron interactions or intracellular processes. Due to their intrinsic voltage sensitivity, they could be used directly as optical readouts of membrane potential. These reporters must be embedded into neural membranes (thickness ~2 nm) and able to react to local electric fields as well as local chemical environments. Functionalization with specific proteins enables the quantum dots to track receptors and functional responses in neurons (*e.g.* to glycine, nerve growth factor, glutamate, *etc.*).

Recent work using tools from atomic physics has shown that optically manipulated color centers in diamond provide exceptionally sensitive magnetic and electric field probes at sub-100 nm distances.[3] Diamond is uniquely suited for studies of biological systems because it is chemically inert, cytocompatible, and ideal for coupling to biological molecules. As a word of caution, it is becoming increasingly apparent that cellular damage may well be a distinct possibility as a consequence of interaction with nanoparticles. Studies of such effects are now included in the field known as nanotoxicology.

4.2 Nanotubes and Nanowires

Nanomaterials that can provide nanoscale topographical features have become popular materials because culture substrates with nanoscale features have significantly different effects on neuronal adhesion and growth. Vertical nanowires were shown to selectively promote neuronal adhesion and guide neurite growth even without any cell-adhesive coating.[4,5] Micropatterned islands of tangled carbon nanotubes also showed similar spontaneous adhesion and growth effects.[6] Guided neuronal growth was reported on various nanotopographical substrates made of nanomesh carbon nanotubes,[7] electrospun nanofibers[8] or patterned polyurethane acrylate.[9]

One-dimensional structures such as nanotubes and nanowires may be used for highly local electrical measurements, for the delivery of photons to specific locations, and for the local release or collection of chemicals. These types of nanodevices could be used alone or combined with conventional organic chromophores, which have been shown to greatly enhance optical signals, hence acting as 'antennae' for light.[10,11] Indeed, membrane-bound and antibody-linked gold nanoparticles have been already used for site-specific measurements of membrane potential. Traditional organic chromophores have suffered from several drawbacks. Due to their large sizes, they can morphologically or chemically perturb the cellular environment. They can bleach, that is, become ineffectual after exposure to light. Nanoparticles, in contrast, can be coated with a passivation layer or specialized shell that limits direct interaction with the surrounding media, hence greatly minimizing bleaching and intracellular generation of reactive oxygen species. Present challenges include developing inorganic nanoparticles with enhanced voltage sensitivity and orchestrating plasmonic enhancement of existing optical reporters. It should also be feasible to develop inorganic nanoparticles with voltage sensitivity, verify their biocompatibility, and identify and validate routes to targeted delivery. New organic non-linear voltage probes and multifunctional nanoprobes that can locally report not only voltage, but also chemical species (*e.g.* calcium, neurotransmitters, *etc.*), local temperature or ionic environment are the focus of current research efforts.

Semiconductor nanowires can detect specific intracellular biomolecules, perform small-molecule drug screening, detect intracellular signaling, and also deliver drugs and genetic material into the cell. This nanotube spearing necessitates an oscillating magnetic field to spear the nanotubes, followed by a static field to drive them inside the cells. Figure 4.2 shows how these nanowires are non-fatal to the cell. The cell remains functional for a few days and can even differentiate from stem cells. The mechanism of cellular engulfment of the nanorods and subsequent normal function remains to be explored.[12]

Silicon nanowires have been implemented with either field effect transistor-type active sensors or metal nanoelectrodes for *in vitro* neural sensors. The

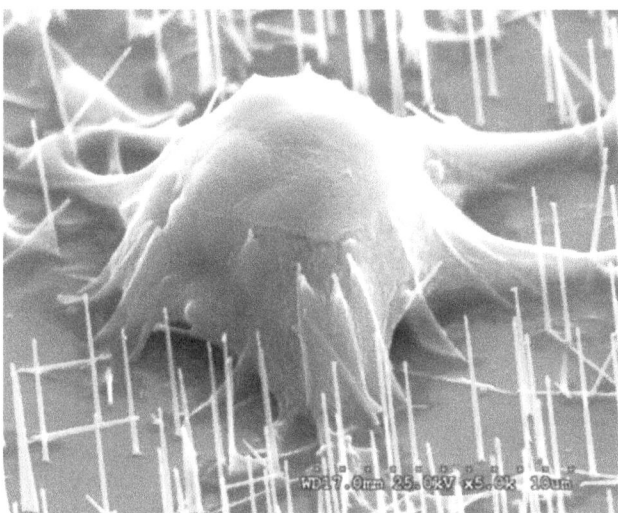

Figure 4.2 Scanning electron microscope (SEM) image of mouse embryonic stem cells penetrated by Si nanowires on a Si substrate. The scale bar represents 10 μm.[12]
(Reprinted by kind permission of the American Chemical Society.)

Lieber group at Harvard University has reported silicon nanowire field-effect transistors (NW-FETs) arrays. They showed that simultaneous recordings from the axon and dendrites of a single neuron were possible with NW-FET arrays.[13] In addition, neural signals ranging from 0.3 mV to 3 mV were recorded from neural circuits in brain slices using a NW-FET array.[14] NW-FET is a promising sensor that can provide sufficient sensitivity with unprecedented spatial selectivity.

Field-effect transistors (FETs) can also record intracellular electric potentials. As their performance does not depend on electrode impedance, they can be made much smaller than micropipettes and microelectrodes. FET arrays are better suited for multiplexed measurements. SiO_2 nanotubes synthetically integrated on top of a nanoscale FET penetrate the cell membrane, bringing the cell cytosol (Figure 4.3) into contact with the FET, which is then able to record the intracellular transmembrane potential.[15]

Branched intracellular nanotube FETs (BIT-FETs) possess a bandwidth high enough to record fast action potentials even when the nanotube diameter is reduced to 3 nm, a length scale well below that accessible with other methods. Studies show that a stable and tight seal forms between the nanotube and cell

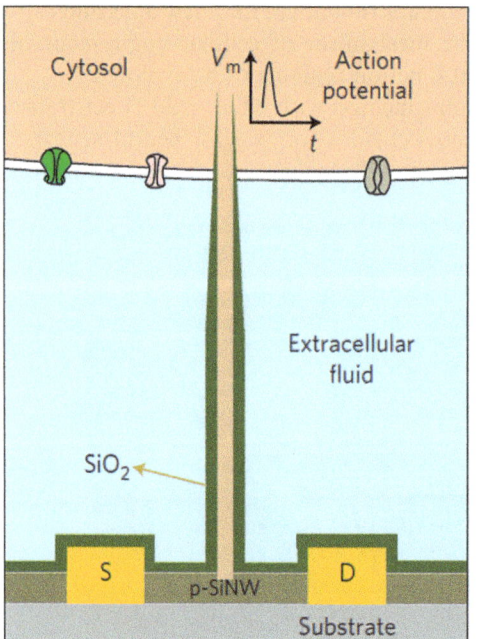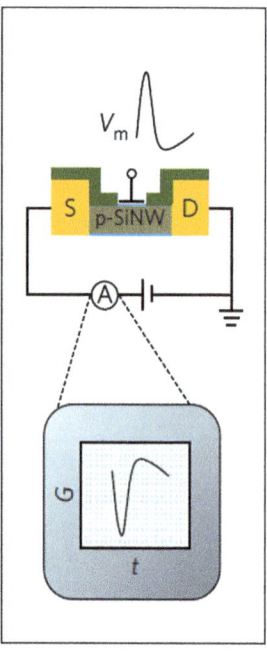

Figure 4.3 Schematic diagrams showing (left) a cell coupled to a BIT-FET and the variation in device conductance G (right) with time, t, during an action potential V_m. S and D indicate source and drain electrodes, respectively. The SiO_2 nanotube connects the cytosol (orange) to the p-type silicon nanowire FET and, together with the SiO_2 passivation (green), excludes the extracellular medium (light blue) from the active device channel.[15] (Reprinted by kind permission from Nature Publishing.)

membrane, leaving the cells undamaged due to the extremely small diameter of the tube. Multiple BIT-FETs can record multiplexed intracellular signals from both single neurons and networks of neurons.

Figure 4.3 shows a nanotube piercing the cell membrane and the FET structure measuring cellular potential changes.

In the case of metal nanoelectrodes, the Park group[15] at Harvard University developed a vertical silicon nanowire array with individual nanowires 150 nm thick and 3 μm high.[16] Several nanowires were grouped (2 μm spacing) to cover a single neuron and an array of grouped nanowires was used to interrogate a small neural circuit. A high signal-to-noise ratio on the order of 100 was achieved with the measured signal amplitude on the order of a few mV.

4.3 Graphene

Ever since the first isolation of freestanding graphene sheets in 2004, this two-dimensional (2D) carbon crystal has been highly anticipated to provide unique and new opportunities for sensor applications. Graphene has already demonstrated great potential in various novel sensors that utilize the exceptional electrical properties of the material (extremely high carrier mobility and capacity), electrochemical properties (high electron transfer rate), optical properties (excellent ability to quench fluorescence), structural properties (one-atom thickness and extremely high surface-to-volume ratio), or its mechanical properties (outstanding robustness and flexibility). Graphene nanostructures exhibiting such excellent properties are very suitable for use as a channel material in FETs as discussed in Chapter 1.

The incorporation of graphene in FETs results in the insertion of a new matrix with superior sensing properties in a structure of high sensitivity, simple device configuration, low cost, high miniaturization and real-time detection. A typical FET consists of a semiconducting channel between two metal electrodes, the drain and source electrodes, through which the current is injected and collected. Varying the gate potential through a thin dielectric layer, typically 300 nm SiO_2, can capacitively modulate the conductance of the channel. In a typical p-type metal oxide semiconductor field-effect transistor (MOSFET), the negative gate potential leads to the accumulation of holes (majority charge carries) resulting in an increase of the channel conductance, while the positive gate potential leads to the depletion of holes and hence a decrease of the conductance. In the case of the electronic sensor, the adsorption of molecules on the surface of the semiconducting channel either changes its local surface potential or directly dopes the channel, resulting in change of the FET conductance. This makes the FET a promising sensing device with easily adaptable configuration, high sensitivity and real-time capability, provided that the issue of non-specific adsorption is addressed by using the appropriate chemistry to prevent the fouling of the surface when the device is exposed to complex biological media such as human serum or blood.

In some cases, the gate electrode is removed to simplify the device structure, hence forming a chemiresistor. Such configuration is suitable for fabrication of graphene-based sensors on polymer substrates for flexible electronics applications. Despite the lack of modulation by the gate potential, the working principle of the chemoresistor is same as a normal FET sensor.

Figure 4.4 shows a typical gas sensing system, in which the channel is directly exposed to a target gas species. The adsorption of gas molecules results in the doping of the semiconducting channel, leading to a conductance change of the FET device. The charge transfer from the adsorbed gas molecules to the semiconducting channel is the dominant mechanism for the current response, which is similar to carbon nanotube based gas sensors.

In order to detect biospecies, the grapheme-FETs should operate in an aqueous environment. As shown in Figure 4(b), the graphene channel is usually immersed in a flow cell or sensing chamber, which is used to confine the solution. The drain and source electrodes are electrically insulated to prevent current leakage from ionic conduction. Different insulators including poly(dimethylsiloxane) (PDMS)/silicone rubber, SiO_2 thin film, SU-8 passivation and silicone rubber are used in different device structures. The gate electrode, usually Ag/AgCl or Pt, is immersed in the solution. The gate potential is applied through the thin electric double layer capacitance formed at the channel–solution interface. The double-layer thickness (or Debye length) is determined by the ionic strength, typically within 1 nm. Normally, the solution-gate FET is over two orders of magnitude more sensitive than the typical backgate FET.

Two major sensing mechanisms have been proposed for graphene-based biosensors in solution, *i.e.* the electrostatic gating effect and the doping effect. The gating effect suggests that the charged molecules adsorbed on graphene act as an additional gating capacitance, which alters the conductance of the graphene channel. However, the doping effect suggests a direct charge transfer between the adsorbed molecules and the grapheme channel, similar to gas sensing. In a real case, the actual sensing mechanism might be a combination of both mechanisms, or involve more complicated mechanisms.

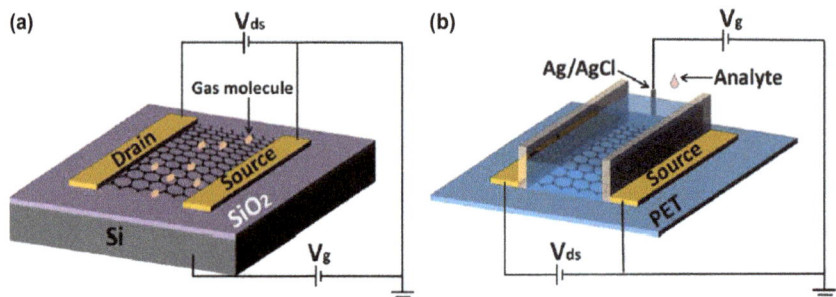

Figure 4.4 (a) Typical back-gate graphene-FET on Si/SiO_2 substrate used as gas sensor. (b) Solution-gate graphene-FET on flexible polyethylene terephthalate (PET) substrate used as chemical and biological sensor in aqueous solution.[17]
(Reprinted by kind permission of the Royal Society of Chemistry.)

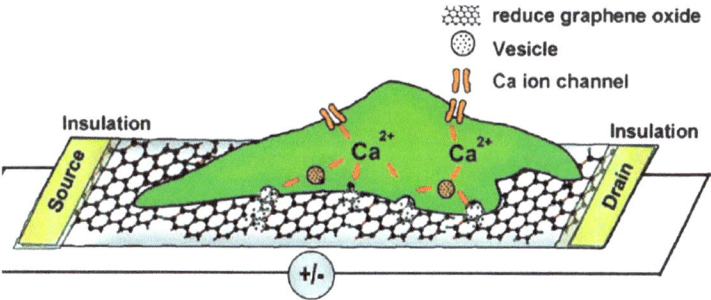

Figure 4.5 Schematic illustration of the interface between a PC12 cell and reduced grapheme oxide FET. The detection was realized by the real-time monitoring of the drain current during the dynamic secretion of catecholamines.[17] (Reprinted by kind permission of the Royal Society of Chemistry.)

The detection of living cells is more challenging as the interaction between the graphene channel and living cell membrane is much more complicated. Graphene offers an improved opportunity to study the cell–nanomaterial interface since its 2D structure provides a homogeneous contact with the 2D cell membrane.

A solution-gate FET geared to investigate living cellular behavior is shown in Figure 4.5. The fabricated devices were based on large-scale micropatterned reduced graphene oxide (rGO) thin films with a thickness of 1–3 nm. Living neuron cells (PC12—a rat pheochromocytoma cell line that acquires many of the characteristics of sympathetic neurons when treated with nerve growth factor) were directly cultured on the rGO channel to obtain an intimate contact. The rGO-FET was able to detect the adsorption of hormonal catecholamine molecules and those secreted from the living PC12 cells with a high signal-to-noise ratio. Moreover, the rGO-FET can be fabricated on the flexible polyethylene terephthalate (PET) substrate and functions well during bending, hence being possibly useful in complicated *in vivo* biosensing.[17]

4.4 Applications of Nanotechnologies in Neuroscience

4.4.1 Nanostructures as Scaffolds for Neuroregeneration and as Interface for Sensing and Stimulation

Carbon nanostructures are potential candidates for incorporation into neural prostheses due to their similar nanoscale dimensions as neurites, as well as their unique electrical and mechanical properties. When being used as a scaffold, they are able to repair injured nerves, and even long gaps in severed nerves, by stimulating the healing of the severed ends in a nerve. Figure 4.6 shows the basic structures used as carbon scaffolds. These structures can be coated with thin layers of polymers in order to decrease the formation of glial scar tissue and to provide suitable sites for cell adhesion and proliferation.[18–20]

Furthermore, the structures can be functionalized in order to improve biocompatibility by decreasing toxicity. Figure 4.7 shows different solutions

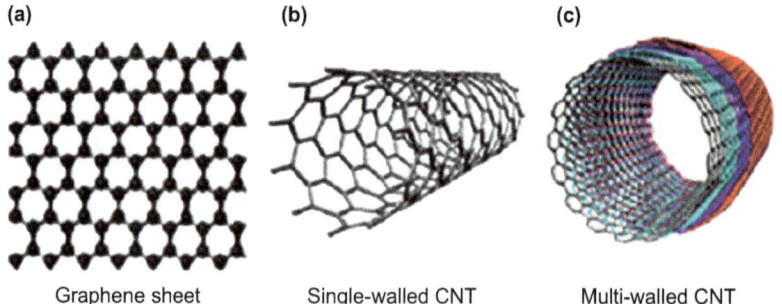

Figure 4.6 Representations of (a) graphene sheet, (b) single-walled carbon nanotube and (c) multi-walled carbon nanotube.[21]

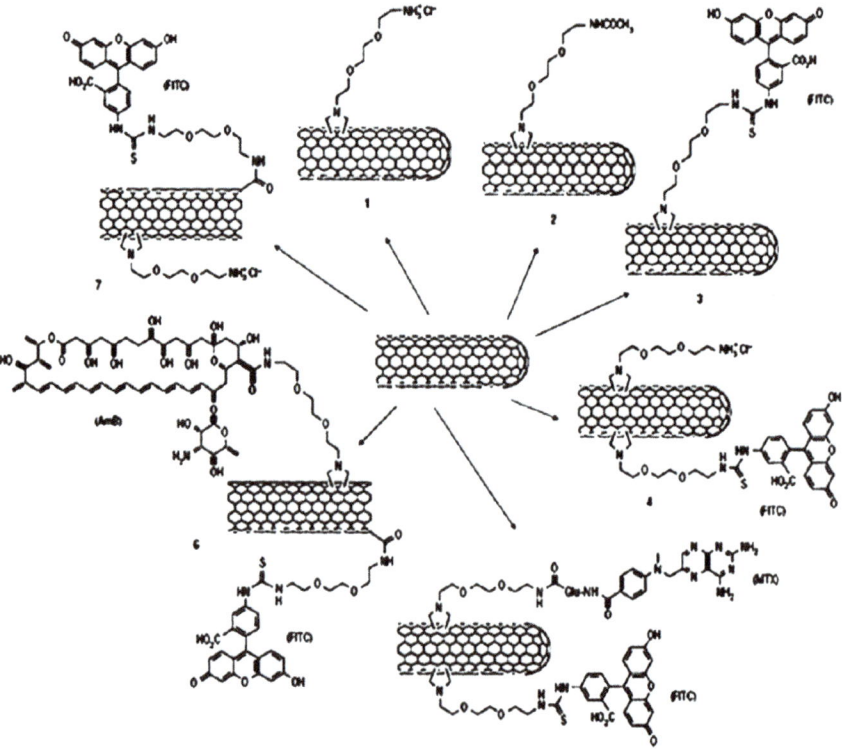

Figure 4.7 Molecular structures of carbon nanotubes (CNT) functionalized covalently with different types of small molecules: 1. Ammonium-functionalized CNT; 2. Acetamidofunctionalized CNT; 3. CNT functionalized with fluorescein isothiocyanate (FITC); 4. CNT bifunctionalized with ammonium groups and FITC; 5. CNT bifunctionalized with methotrexate (MTX) and FITC; 6. Shortened CNT bifunctionalized with amphotericin B (AmB) and FITC; 7. Shortened CNT bifunctionalized with ammonium groups and FITC (through an amide linkage).[21]
(Reprinted by kind permission of Elsevier BV.)

implemented in order to explore the controversial aspect of carbon nanotube toxicity, which manifests itself due to carbon nanotubes blocking ion channels in the membrane, causing increased cell oxidative stress, reduced cell adhesion, or apoptosis.

A recent improvement in such prosthetic medical devices involved growing carbon nanotubes on a biocompatible Pt catalyst.[21] The new material shows less cellular degeneration due to oxidative stress as the main product of the Pt catalyzed oxygen reduction is water. This is ideal for guiding the axon regeneration through tubes made of materials such as chitosan, a biocompatible and biodegradable natural material that can provide, as a gel sponge, a suitable scaffold for nerve regeneration.

4.4.2 Nanoribbons for Sensing Cellular Deformation

Although the electrical response of neurons to applied voltage has been studied extensively, mechanical response has been largely ignored, even if it might advance the studies on cellular function and physiology, especially in the area of axon elongation and dendritic formation. Using piezoelectric nanoribbons made of PbZr or $Ti_{1-x}O_3$, it was found that the cells deflect by 1 nm when 120 mV is applied to the cell membrane. Such depolarization induces changes in the membrane tension so that it is accommodating the stimulus by equalizing the overall pressure across the membrane through a process resembling converse flexoelectricity.[22]

Figure 4.8 shows piezoelectric nanoribbons suspended over a trench as nanobeams to maximize deflection. The use of an underlying substrate of transparent MgO as well as transparent indium tin oxide (ITO) electrodes facilitates backside chip visualization during electrophysiology measurements. The electrodes are electrically isolated by a coating of SiN_x to ensure no cross-signal response when the chip is placed into solution. PC12 cells were cultured on the piezoelectric chip, and those cells located on the nanobeam arrays were patch-clamped with a standard glass electrode for membrane voltage stimulation.

4.5 Challenges and Future Perspective of Nanotechnologies in Neuroscience

It is not clear how neurons interact with nanostructures, why they continue to function when impaled by nanospears from all directions and how they heal when confined in nanoscaffolds. Reports on the toxicity of nanomaterials are extremely controversial. However, if any such undesirable effect can be minimized and these new materials can contribute to brain regeneration and repair, the outcome might be a positive one for the future. Like the miniaturized submarine crew in the 1966 movie *Fantastic Voyage*, smart nanodevices may one day be sent beyond the blood–brain barrier to perform lifesaving surgery inside the brain tissue and to destroy cancerous brain tumors with extreme precision and in a minimally invasive mode, by delivering the necessary drug and then leaving the body in a harmless way. It would represent a less

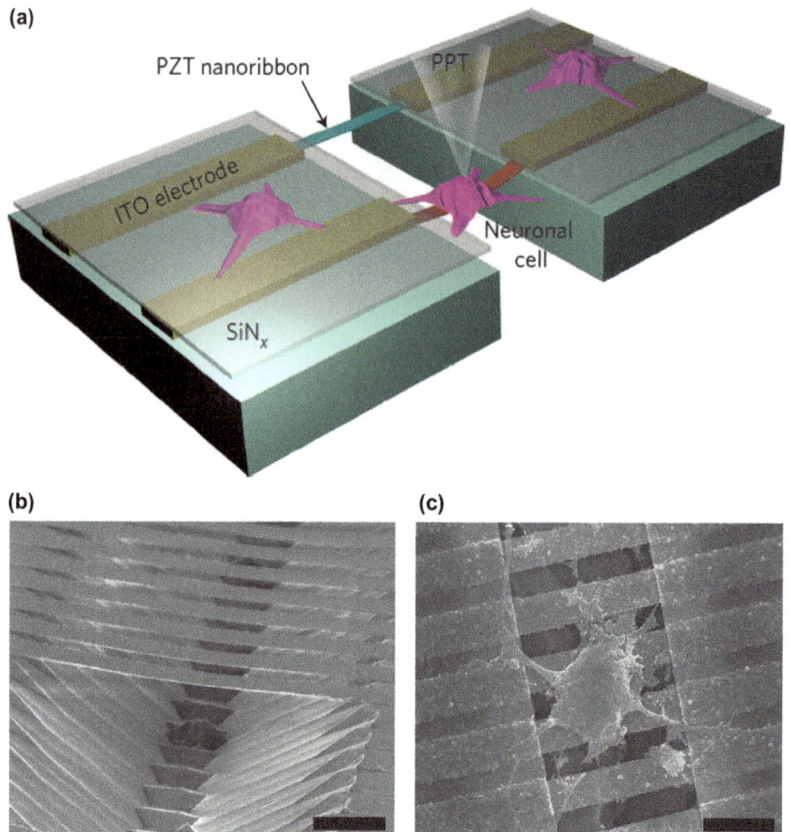

Figure 4.8 Interfacing of PZT nanoribbons with cultured neuronal cells. (a) Schematic of the piezoelectric nanoribbon device with cultured neuronal cells. The suspended nanoribbons record cellular mechanical deflections while the glass pipette (PPT) applies and records membrane potentials. (b) SEM image of suspended PZT nanoribbons (scale bar, 5 mm). (c) SEM image of a single PC12 cell directly interfaced with suspended PZT nanoribbons (scale bar, 15 mm).[22]
(Reprinted by kind permission of Nature Publishing.)

expensive and far more effective treatment because only the target cells will be affected, without the side effects of today's chemotherapy. Such an approach to medicine has the potential to transform healthcare for everybody.

References

1. G. E. Moore, Cramming more components onto integrated circuits, *Electronics Magazine*, 19 April 1965.
2. S. Pathak, E. Cao, M. C. Davidson, S. Jin and G. A. Silva, *J. Neurosci.*, 2006, **26**, 1893.

3. S. Guo, S. D. Solares, V. Mochalin, I. Neitzel, Y. Gogotsi, S. V. Kalinin and S. Jesse, *Small*, **8**, 1264.
4. A. K. Shalek, J. T. Robinson, E. S. Karp, J. S. Lee, D. R. Ahn, M. H. Yoon, A. Sutton, M. Jorgolli, R. S. Gertner, T. S. Gujral, G. MacBeath, E. G. Yang and H. Park, *Proc. Natl. Acad. Sci. U. S. A.*, **107**, 1870.
5. W. Hallstrom, T. Martensson, C. Prinz, P. Gustavsson, L. Montelius, L. Samuelson and M. Kanje, *Nano Lett.*, 2007, **7**, 2960.
6. R. Sorkin, A. Greenbaum, M. David-Pur, S. Anava, A. Ayali, E. Ben-Jacob and Y. Hanein, *Nanotechnology*, 2009, **20**, 015101.
7. M. J. Jang, S. Namgung, S. Hong and Y. Nam, *Nanotechnology*, **21**, 235102.
8. C. C. Gertz, M. K. Leach, L. K. Birrell, D. C. Martin, E. L. Feldman and J. M. Corey, *Dev. Neurobiol.*, **70**, 589.
9. K. J. Jang, M. S. Kim, D. Feltrin, N. L. Jeon, K. Y. Suh and O. Pertz, *PLoS One*, **5**, e15966.
10. P. L. Stiles, J. A. Dieringer, N. C. Shah and R. P. Van Duyne, *Annu. Rev. Anal. Chem.*, 2008, **1**, 601.
11. F. Tam, G. P. Goodrich, B. R. Johnson and N. J. Halas, *Nano Lett.*, 2007, **7**, 496.
12. S. J. Pearton, T. Lele, Y. Tseng and F. Ren, *Trends Biotechnol.*, 2007, **25**, 481.
13. F. Patolsky, B. P. Timko, G. Yu, Y. Fang, A. B. Greytak, G. Zheng and C. M. Lieber, *Science*, 2006, **313**, 1100.
14. Q. Qing, S. K. Pal, B. Tian, X. Duan, B. P. Timko, T. Cohen-Karni, V. N. Murthy and C. M. Lieber, *Proc. Natl. Acad. Sci. U. S. A.*, **107**, 1882.
15. X. Duan, R. Gao, P. Xie, T. Cohen-Karni, Q. Qing, H. S. Choe, B. Tian, X. Jiang and C. M. Lieber, *Nat. Nanotechnol.*, **7**, 174.
16. J. T. Robinson, M. Jorgolli, A. K. Shalek, M. H. Yoon, R. S. Gertner and H. Park, *Nat. Nanotechnol.*, **7**, 180.
17. Q. He, Z. Zeng, Z. Yin, H. Li, S. Wu, X. Huang and H. Zhang, *Small*, **8**, 2994.
18. A. Yokoyama, Y. Sato, Y. Nodasaka, S. Yamamoto, T. Kawasaki, M. Shindoh, T. Kohgo, T. Akasaka, M. Uo, F. Watari and K. Tohji, *Nano Lett.*, 2005, **5**, 157.
19. P. A. Tran, L. Zhang and T. J. Webster, *Adv. Drug Delivery Rev.*, 2009, **61**, 1097.
20. M. Foldvari and M. Bagonluri, *Nanomedicine*, 2008, **4**, 173.
21. F. Tavangarian, Y. Li, *Ceramics International*, 2012, **38**, 6075.
22. T. D. Nguyen, N. Deshmukh, J. M. Nagarah, T. Kramer, P. K. Purohit, M. J. Berry and M. C. McAlpine, *Nat. Nanotechnol.*, **7**, 587.

CHAPTER 5
The Vibrational Field and Detection of Neuron Behavior

5.1 Extending Human Sensory Capabilities

Sensors are intricate devices able to extend the limited brain capabilities in the information exchange with the environment. Our five senses—sight, hearing, smell, taste and touch—are physiological sensory channels acquiring only the narrow input necessary for the experiences and reactions we identify as survival. We need to build microscopes and telescopes, infrared and ultraviolet detectors because the human eyes perceive only the narrow electromagnetic radiation from 400 to 700 nm in wavelength, oblivious to all the other frequencies of the spectrum. Likewise, the human ear responds to a very narrow range of acoustic frequencies, even narrower after childhood and sometimes needing hearing aids later in life. We are insensible to the indiscernible infrasound and ultrasonic realms of acoustic oscillation. So we assemble seismographs, radio and microwave technologies to extend far beyond our restricted capabilities of sensing.

There is also the so-called sixth-sense, which refers to intuition, instinct and inspiration, something that each of us experience every day. However, we cannot create sensors for it simply because we do not know how to directly measure something so non-material with material instruments. Since mind and consciousness are difficult to comprehend, still considered emerging from the physiological brain as epiphenomena of its biochemical states, we will focus our discussion on how biophysical devices can detect neurological events at cellular and molecular level. The hypothesis that the brain is a transducing 'funnel' (as Aldous Huxley called it in *The Doors of Perception*) for the universal conscience belongs to the fluid frontiers of human knowledge, where the unexplained must still be explored. We feel that our mind is creating the meaning of all our sensory, mental and emotional experiences through the nervous system, while consciousness seems to transcend space and time, not being bound to the brain activity.

In all living cells, biochemical and molecular reactions are fundamentally electrical interactions. Using electromagnetic fields of different frequencies to probe the structure and function of whole single cells or entire populations of cells is a non-invasive, and integrative method for understanding subtle cellular mechanisms and offers valuable information on the complex behavior of living cells. Cellular biosensors based on innovative novel approaches in electrical impedance spectroscopy, scanning Kelvin microscopy as well as in vibrational acoustical and optical techniques constitute new, label-free and non-invasive platforms for measuring the transduction and processing of cellular signals. Since each cell can be modeled as a miniature resonant electrical circuit, slight changes in cytoplasm resistivity and membrane capacity can be detected and evaluated in parallel to classical bioanalytical assays.

Energy vibrations of different wavelengths can discern intricate relaxation processes originating in the reorientation of dipole moments of biomolecules in alternate fields and in the variation of dielectric properties of cells and their different constituents. Work function variations induced by chemical perturbation can reflect alterations in cell behavior at submolecular level due to compounded fluctuations of electrical charges and dipolar rotations. The applications are innumerable, ranging from fundamental studies on morphological and structural changes that occur under different physical, chemical, pharmacological and biological stimuli or under specific micro-environmental conditions, to detecting ionic mechanisms such as ionic channels opening, immunological relevant cell modifications by antigen–antibody interactions taking place on the cell surface, measuring cell viability and monitoring cell growth, discrimination of leucocytes and red blood cells, discrimination of normal and malignant cells, stem cells differentiation, cell–cell communication in neuron cultures grown on the surface of the biosensors, and tissue regeneration.

5.2 The Vibrational Field as a Neural Sensor Platform

Detection methods applied in cellular biology are, by their nature, extremely interdisciplinary in character. This is especially the case for cellular biosensors which are based on the work of highly multidisciplinary teams that include individuals with the passion and scientific curiosity to investigate apparently diverse sciences such as physics, chemistry, biology and medicine, and also engineering, instrumentation, microsystem technologies, microelectronics and so on. A key feature of research in this area is the bridging of gaps formed by traditional specialization of modern scientific endeavors. Using vibrational fields to detect molecular behavior and cellular events represents the state-of-the-art in the field of modern detection science. Electromagnetic and acoustic fields of variable frequencies are perfectly non-invasive and label-free techniques to monitor and record in real time cell characteristics, structural changes and function. Unlike patch-clamp techniques, such methods can offer continuous measurements while the cells are stimulated, without destroying the cells in the process. Observations can be made after the stimulus is removed, on a still living cell. Bioelectronic devices based on fields of variable frequency can measure cellular activity and interaction with drugs, leading to spectacular applications ranging from nanomedicine to repair and regeneration technologies. Today, classical microscopy (optical, electron microscopy), which shows only the physical aspect of the cell, is plagued by detection limits. It can be complemented by high-resolution, extremely sensitive methods translating the spatial and temporal characteristics of living systems into the frequencies and amplitudes of a complex transducing signal, carrying ample and valuable information. Such methods have the potential to open new windows in cell biology and neuroscience in particular.

5.2.1 The Simple Vibrating Probe

The history of the scientific development of cellular biosensors is extremely relevant to the current state-of-the-art. In the 1970s Richard Nuccitelli discovered that small direct current (DC) electric signals (having important functions in processes such as embryo development, cell migration, wound healing, *etc.*) can be detected around living cells using a small vibrating probe.[1] The probe consisted of a metal wire 10–30 μm in diameter that was sharpened, had a small platinum black tip and was electrically insulated. It was set in vibration at 300 Hz by a piezoelectric bender (Figure 5.1).

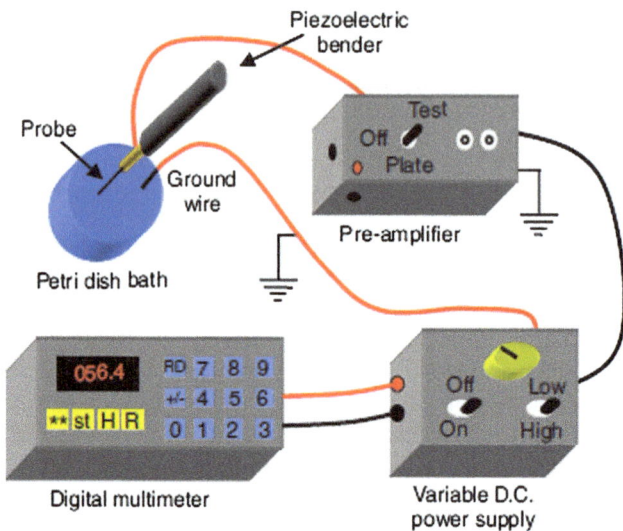

Figure 5.1 Set-up of vibrating probe technique for measuring cellular currents.[1] (Reprinted by kind permission of the Rockefeller University Press.)

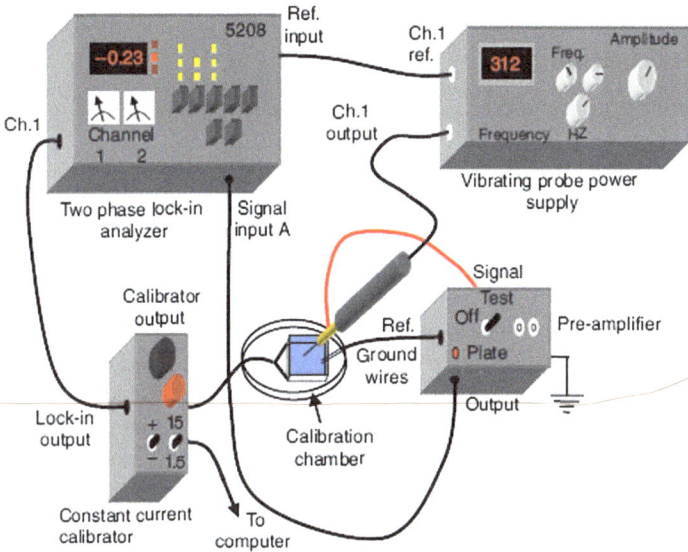

Figure 5.2 Calibration procedure for vibrating probe.[1]
(Reprinted by kind permission of the Rockefeller University Press.)

In the presence of a physiological ionic current, the probe detects a voltage difference between the extremes of its position. The calibration procedure is rather complicated (Figure 5.2), but the system is low-cost and useful in detecting bioelectric currents generated by living cells. For example, Nuccitelli has shown that a steady current with an average surface density of 0.1–0.2 μA cm^{-2} enters the tail of an amoeba and leaves at its pseudopodes. This current is reduced when the Ca^{2+} in the media is reduced. It is either a current carried by Ca^{2+} ions or the absence of calcium affects the fluxes of other ions such as Cl^-. Jaffe later demonstrated the detection of subtle calcium gradients using calcium selective vibrating electrodes in different embryonic systems during cellular development. The method can be extended to measure any ionic gradients of interest (H^+, Mg^{2+}, K^+, Na^+).[2] In a surprising twist of fate, Nuccitelli's work is now resurrected and was recently published in leading scientific journals.[3]

One interesting practical application of this technique is the non-invasive measurement of electric current in human corneal wounds using a vibrating probe to test the feasibility of pharmacologically enhancing the current to promote corneal wound healing. Unwounded cornea were found to have small outward currents (0.07 μA cm^{-2}).[4] Wounding increased the current more than five-fold (0.41 μA cm^{-2}). Monitoring the wound current over time showed that it seemed to be actively regulated and maintained above normal unwounded levels for at least six hours. Drug treatment with aminophylline or chloride-free solution more than doubled the size of wound currents, which correlated directly with the wound healing rate.

5.2.2 Electric Impedance Sensing of the Cell–Substrate Interaction

The general approach used in this methodology was introduced in Chapter 1. In this case, instead of the placement of a vibrating electrode in the vicinity of cells, the latter can be deposited on a substrate where a very small electrode is surrounded by a larger electrode structure (Figure 5.3). Impedance spectroscopic measurements can then be performed at variable frequencies.

Such a structure, excited at 4 KHz frequency, was designed by Ivar Giaever, Nobel Laureate in Physics, who now works in biophysics. The key for such measurements is the dimension of the working electrode. Unless the electrode is extremely small, the impedance of the culture medium, in series with impedance of the electrodes, will dominate the measurement. The constriction resistance of the solution is directly proportional to the resistivity and inversely proportional to the diameter of the electrode. This can be overcome by the electrolyte–electrode interface which is inversely proportional to the square of electrode diameter. If the electrode diameter is 10^{-3} cm and the frequency is chosen at around 4 kHz, the Faradaic resistance is several times larger than the constriction resistance. In this way, the activity of the cells anchored to the electrode can be revealed. If two large electrodes had been used instead, the solution resistance would have concealed the signals. The principle at the core of this method is illustrated in Figure 5.4.[5] The electric cell–substrate impedance sensing system (ECIS) is commercially available and currently used in applications such as study of attachment and spreading of living cells, cellular growth and proliferation, angiogenesis, wound healing, cell migration, *etc.*[6–14]

5.2.3 Miniaturization of the Electrical Impedance Tomography Technique

In the electric impedance technique, if the electrodes are arranged radially around the cell, impedance measurements can be used to map the membrane

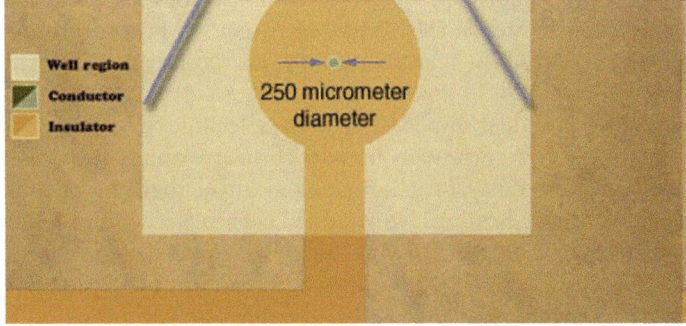

Figure 5.3 Impedance spectroscopy: electrodes set up for measuring the metabolism of living cells attached to the substrate.[5]
(Reprinted by kind permission of Nature Publishing Group.)

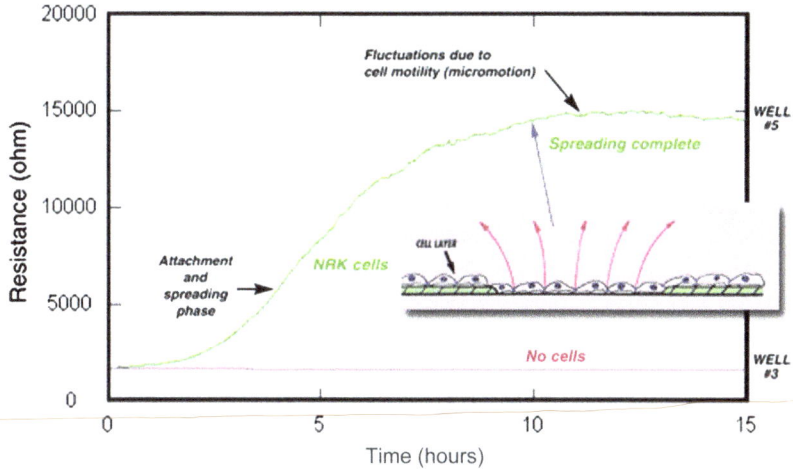

Figure 5.4 The principle of the ECIS system.[6]
(Reprinted by kind permission of Nature Publishing Group.)

capacity through detection of the spatial distribution of the displacement current (Figure 5.5). Cellular interrogation takes place at frequencies between 10 kHz and 1 MHz. The system can be adapted to work under a microscope and also facilitates the applications of chemical solutions by microinjection. Applications range from synaptic transmission, ion channel activity, cell division and growth to protein-based electromotility and membrane flexoelectricity.[15] The technique is essentially a miniaturization of electrical impedance tomography, which is an imaging method that uses a harness of electrodes around the human torso, for example, to visualize the organs based on the specific impedance of each human tissue. Unlike the situation where ionizing radiation is employed, such as involved with X-ray imaging, impedance tomography is a much gentler method and can be applied without serious, medical side effects.

An integrated cell biochip microsystem platform based on bioimpedance measurements has been developed at the Tyndall National Institute in Ireland. The platform combines mixed sensor technologies with microfluidics capabilities, enabling multi-parameter detection.[16–18]

Optical imaging and electrical (impedance) detection, pH, temperature and dissolved oxygen sensors and microfluidics are all integrated into the system. Impedance measurements are performed on inter-digitated electrode structures (IDES) of 20 µm width and 10 µm gaps, fabricated using indium tin oxide (ITO) chosen as biosensor material due to its optical transparency, electrical characteristics and biocompatibility. The fluidic capabilities allow cell media to be delivered to each of the system's six modules, enabling chemicals to be delivered to the cells. Concentration, growth and alterations of the physiological state of cells can be detected as impedance changes. The system yields information about spreading, attachment and morphology of cells. Figure 5.6 shows an image of an ITO biochip surface plated with live cells. The effect of pharmaceutical drugs and neurotoxins can be studied. The technology uses

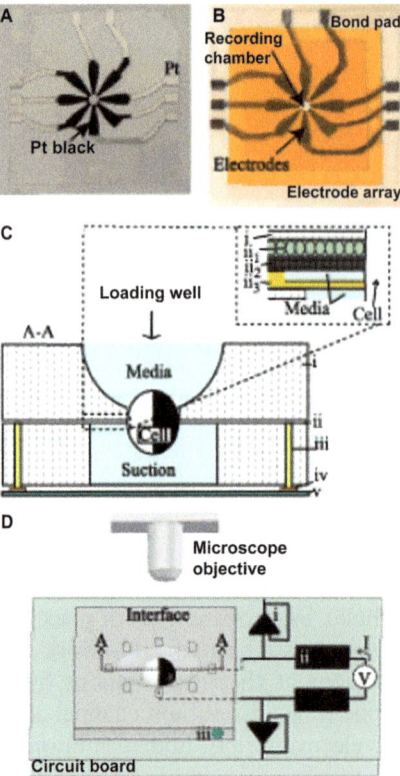

Figure 5.5 Single cell impedance measurements set-up.[15] A) Platinum was patterned onto a polyester substrate, and the electrodes near the recording chamber were plated with platinum black. B) Insulated, electrode array. C) A polycarbonate interface (C-i) allowed for positioning of the electrode array (C-ii) such that spring-loaded gold pins (C-iii) connected the bond pads on the electrode array to bond pads (C-iv) on a printed circuit board (PCB, C-v). Inset shows the multiple layers of the electrode array (a polyester base (ii1), recessed platinum-black coated electrodes (ii2), and insulating Kapton tape (ii3)). D) PCB and fluidic interface positioned on the stage of an upright microscope showing the headstage amplifiers (D-i), computer controlled digital switches (D-ii) and vacuum-port (D-iii). (Reprinted by kind permission of Royal Society of Chemistry Publishing.)

impedance analysis which enables accurate, sensitive and reliable assays to be performed in real time and under constant automated monitoring. These types of biosensors also offer the potential to study the behavior of neuron cells in a non-destructive assay format which may provide considerable benefit to those working in research areas such as neurology, cytotoxicity and pharmacology. It addresses markets such as the pharmaceutical industry, environmental monitoring, healthcare, and security/defense sectors.

An interesting development with respect to impedance-based techniques is their application to the study of neurodegenerative diseases as illustrated by the work of the Sierks group at Arizona State University, USA. The focus of this

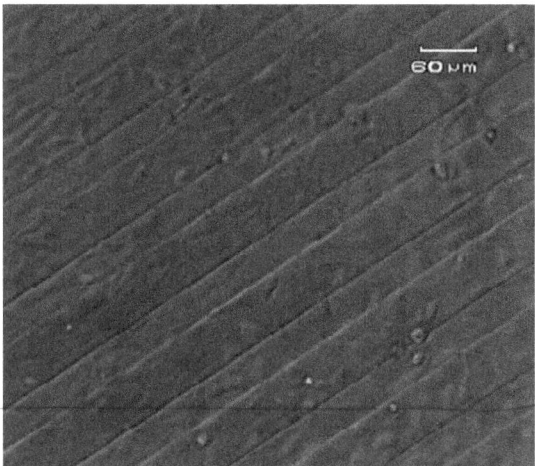

Figure 5.6 Cells growing on the biochip surface.[16]
(Reprinted by kind permission of Dr Paul Galvin, Tyndall National Institute Cork City, Republic of Ireland.)

type of research is the detection of misfolded and aggregated protein variants that contribute to many human diseases including Alzheimer's disease (AD), Lewy body dementia (LBD), frontotemporal dementia and Parkinson's disease (PD). The diagnosis of such diseases is very challenging.

Cerebral spinal fluid (CSF) levels of Aβ, tau and phosphorylated tau are promising biomarkers for diagnosing AD. There is increasing evidence to indicate that various soluble, aggregated oligomeric forms of Aβ, a-syn and tau are the relevant toxic species in different neurodegenerative diseases, and specific detection of different aggregate species in CSF may provide a more refined and powerful tool to facilitate early and accurate diagnosis of a variety of such medical conditions. Additionally, this type of detection approach may lead to an understanding of the mechanisms involved in the onset and progression of these diseases. Protein aggregation is a common thread behind numerous neurodegenerative diseases including AD, PD, LBD, tauopathies and synucleinopathies. Aggregation of Aβ has been correlated with AD, aggregation of a-syn with PD, LBD and other synucleinopathies, and aggregation of tau with AD and various tauopathies.

While the presence of fibrillar aggregates of these different proteins has been a classic diagnostic feature of the respective diseases, increasing evidence suggests that soluble oligomeric forms of these proteins are the relevant toxic species. During the polymerization process from monomeric to fibrillar form, each of the protein species must pass through different oligomeric states, suggesting that various oligomeric species may represent earlier biomarkers for these diseases compared with the presence of fibrillar forms. A rapidly growing body of evidence indicates that oligomeric forms of Aβ are key factors in the onset and progression of AD. Aβ forms a number of soluble intermediate or metastable structures which may contribute to toxicity. Cortical levels of soluble Aβ correlate well with the cognitive impairment and loss of synaptic function. Small, soluble spherical or annular aggregates of Aβ were shown to be neurotoxic and oligomeric forms of Aβ, created *in vitro* or derived from cell cultures inhibit long-term potentiation (LTP).[19] The concentration of oligomeric forms of Aβ are elevated in transgenic mouse models of AD, human AD brain and CSF samples, and the presence of stable dimeric form of Aβ associates well with dementia in AD patients. Disruption of neural connections near Aβ plaques was also attributed to oligomeric Aβ species, a halo of oligomeric Aβ surrounds Aβ plaques causing synapse loss and oligomeric Aβ was shown to disrupt cognitive function in transgenic animal models of AD. Different size oligomers of Aβ have been correlated with AD, including a 56 kD aggregate and smaller dimeric, trimeric and tetrameric species.[19] Therefore specific detection of soluble oligomeric Aβ species holds great promise as a biomarker for studying the progression of AD. Similarly, formation of oligomeric aggregates of a-syn has also been correlated with PD and other synucleinopathies. A-syn is a major component of Lewy bodies and neurites. Aggregated forms of a-syn were shown to induce toxicity in dopaminergic neurons *in vivo* and several different oligomeric morphologies were shown to each have different toxic mechanisms and effects on cells.[19] The presence of various oligomeric a-syn species in CSF is also a very promising biomarker for studying the progression of various neurodegenerative diseases.

Detection of small toxic oligomeric aggregates of proteins such as β-amyloid or α-synuclein in human CSF or serum samples may provide a means to study disease mechanisms and progression, and a means to presymtomatically diagnose these diseases. To detect the presence of different protein variants in CSF, the use of novel reagents that selectively recognize specific protein isoforms is coupled with a novel electronic biosensor that has femtomolar sensitivity and is compatible with untreated human samples. The presence of specific oligomeric protein species in human CSF samples can readily distinguish between different neurodegenerative diseases and this technology has potential to facilitate early diagnosis of these diseases.[19]

Figure 5.7 illustrates the biosensing platform, which is comprised of three integrated parts: (a) a printed circuit board (PCB); (b) a nanoporous alumina membrane; and (c) a silicone microfluidic chamber. The PCB platform consists of inter-digitated working and counter electrodes. The tin oxide electrodes are 800 mm in width, 5 mm in length and 800 nm in thickness with rounded edges to minimize fringe effects during the application of a sinusoidal voltage input signal (Figure 5.7(a)). A nanoporous alumina membrane is soldered onto the inter-digitated electrodes generating a high density array of nanowells. The membrane is 250 nm thick, has a lateral diameter of 13 mm with pore diameters of 200 nm. The porosity of the membranes varies between 25% and 50%. An alumina membrane was utilized since it offers electrical isolation between each individual nanowell as well as good biocompatibility (Figure 5.7(b)). Finally, a circular silicone chamber encloses the nanotextured electrode surface to confine the fluid onto the device surface and prevent evaporation, which could lead to electrical signal instability. The chamber has a maximum working volume of 1.6 mL. Thus, the combination of the alumina membrane on a substrate

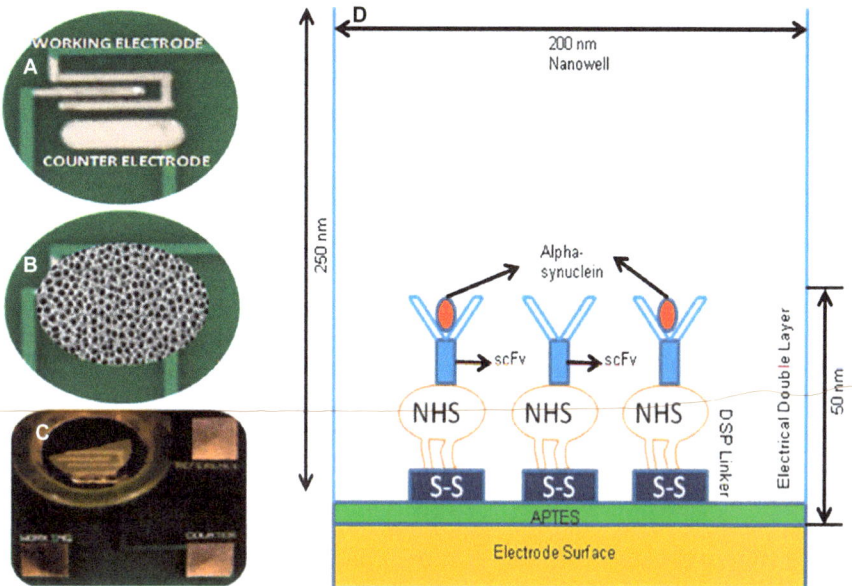

Figure 5.7 Impedance biosensing platform: (A) printed circuit board; (B) nanoporous alumina membrane; and (C) silicone microfluidic chamber. (D) Changes to the electrical double layer within each nanowell due to the binding of the target antigens onto the nanobodies. Nanobodies are immobilized onto the electrode sensor surface using a chemical linker. The electrode surface is first amine functionalized using 3-aminopropyl triethoxysilane (APTES). The alumina membrane is then soldered to the silanized (S–S) electrode surface. Then 3,3′-dithiobis succinimidyl propionate (DSP) is used to cross link the nanobodies to the electrode surface using N-hydroxysuccnimide (NHS). The DSP (thiol linker) is to allow conjugation to the silanized electrode surfaces, which form the base of the nanowells. After conjugation of the linker to the nanowell surfaces, an aliquot of the nanobody (scFv single chain variable fragment shown) is added, followed by addition of bovine serum albumin (BSA) to block any unbound amine sites on the sensor surface.[19]
(Reprinted by kind permission of the Royal Society of Chemistry.)

enclosed by a silicone chamber forms an inexpensive biosensing device capable of detecting various biomolecules (Figure 5.7(c)).

The biosensor measures impedance changes to the electrical double layer at the solid–liquid interface within the nanowells induced when target proteins contained in the sample bind to reagents such as antibodies immobilized on the sensor surface. Impedance measurements provide very detailed information about the electrical changes occurring at the interfaces. When target antigens bind immobilized antibody inside the nanowells, the double layer capacitance changes due to the change in the surface charge concentrations. Thus, the double layer capacitance directly correlates to the amount of binding taking place at the solid–liquid interface and the amount of binding is directly proportional to the concentration of the target species. Thus, by characterizing the double layer capacitance, an accurate estimate of the concentration of the target species can be measured. The changes to the double layer capacitance can be represented as the measured impedance changes especially at low frequencies (below 1 kHz). In the implemented sensor configuration, redox probes are not used and it can be assumed that all the conduction occurring at the interface is non-faradaic in nature, so the charge distribution dynamics at the metal–solution interface characterizing the biomolecular interactions at the surface can be modeled using the Helmholtz–Gouy–Chapman model with Sterns correction. Since binding of antigens to the nanobodies is free of any biochemical mediators, the impedance changes within the electrical double layer (Figure 5.7(d)) is non-Faradaic and the electrical circuit model of the sensor can be represented as a simple resistive–capacitive (RC) series circuit whose values are extracted by a frequency response analyzer potentiostat. For probing the impedance changes to the electrical double layer of the nanowell electrodes, a very small amplitude sinusoidal voltage is applied to the electrochemical system and the output current response is sensed. The ratio of the applied voltage phasor to the output current phasor is the resulting impedance, which is characterized using a frequency response analyzer.

A commonly used electrochemical immunosensing technique is pulsed amperometry. This method involves the immobilization of an immune-reagent component on the electrode transducer and the use of an electrochemical active substance produced by enzymatic reaction for signal generation. As simple as this appears, there can be numerous problems associated such as inadequate supply of enzyme inhibitors in the sample, instability of the enzyme over time, irreproducibility of the electrode kinetics for the re-oxidizing reagent or reducing oxidizing agent, redox active interferences which either react at the electrode and/or couple with the reagent couple, and inadequate temperature control. The electrochemical impedance spectroscopy (EIS) technique eliminates most of these problems since it doesn't rely on the redox properties of the analyte and doesn't need an enzyme inhibitor. Another fundamental difference between the two techniques is the sensing mechanism. Amperometry involves detection of ions in the solution by applying a fixed voltage through electrodes and measuring the current/change in current, whereas EIS involves characterizing the electrical double layer at the electrode by sweeping a range of frequencies and measuring the current.

Highly selective morphology specific reagents (isolated nanobodies) are immobilized onto the electrode sensor surface using a chemical linker. The electrode surface is first amine functionalized using 3-aminopropyl triethoxysilane (APTES, 2% in acetone buffer). A 100 mL aliquot of 2% APTES is applied on the electrode surface and incubated at room temperature for 30 s. Excess APTES is then removed by flowing acetone over the surface. The alumina membrane is then soldered to the silanized electrode surface. Then 3,3′-dithiobis succinimidyl propionate (DTSSP) dissolved in dimethyl sulfoxide (DMSO) solvent (4 mg mL^{-1}) is used to cross-link the nanobodies to the electrode surface.

Specific aggregate morphologies of a-syn and Aβ are likely to be present in CSF samples only at very low concentrations (nanomolar or less) and therefore successful detection of these targets requires a biosensor with very low detection limits. The first step to determine biosensor sensitivity is to determine suitable electrical parameters which will enable detection of the bound target antigen using electrochemical impedance spectroscopy. Since binding of antigen to target takes place over a high density array of nanopores, the impedance signal correlates to the average signal obtained over all the pores, ensuring that even if some pores do not contain immobilized capture agent the measured impedance will be reproducible within an acceptable margin of error. The dimensions of the electrical double layer within the nanopores are approximately 50 nm. An 100 mV peak-to-peak pulse is utilized to characterize changes to the capacitance of the electrical double layer induced by biomolecules binding. The second parameter that needs to be defined is the sensing frequency. Since double layer capacitance dominates the impedance spectrum for frequencies less than 1 kHz, in order to determine the optimum frequency for these studies, the frequency response is recorded by adding a range of monomeric target antigen concentrations to immobilized antibody using a frequency range from 50 Hz to 1 MHz. A frequency of 100 Hz gave maximum visible shifts in the impedance induced by biomolecule binding (Figure 5.8).[19] Using the defined

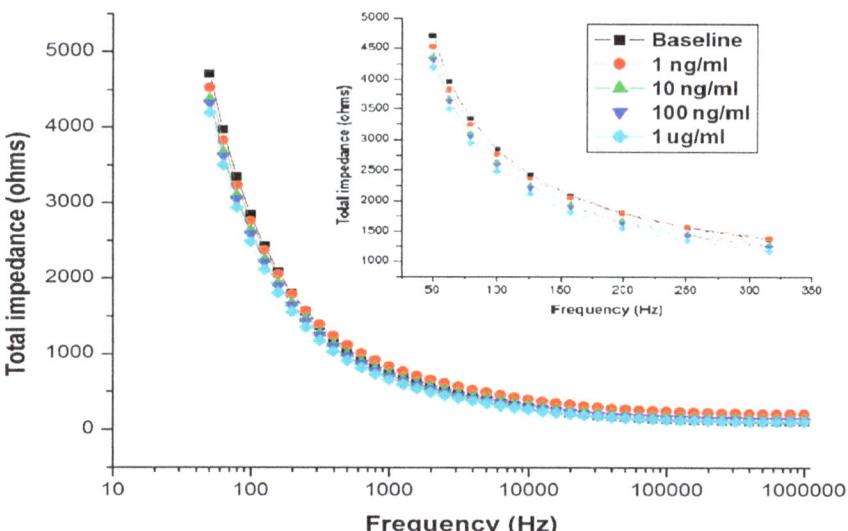

Figure 5.8 Choosing the most discriminating frequency value for maximum sensitivity detection from a range from 50 Hz to 1 MHz: the inset shows a zoomed spectrum for various doses at relevant frequencies. Because 100 Hz showed maximum separation shifts between the spectra for individual doses in the impedance changes induced by biomolecular binding, it was chosen as the sensing frequency.[19]
(Reprinted by kind permission of the Royal Society of Chemistry.)

electrical parameters, a calibration curve was generated using different concentrations of target antigen molecule using two controls; bovine serum albumin (BSA) was added to the immobilized antibody to test for nonspecific binding to nanobody, and target antigen without immobilized antibody was used to measure background binding to the sensor surface. From the calibration curve, it is possible to accurately detect antigen down to a limit of detection of 1 picogram mL^{-1} indicating that it should be possible to detect low femtomolar concentrations of target antigen in clinical patient samples.

These studies suggest that label-free detection of specific oligomeric aggregate species holds great promise for use as sensitive biomarkers for neurodegenerative disease.

5.2.4 Optical Sensing Platforms

Optical biosensors based on whole cell sensing are gaining widespread use in cellular research largely as a result of a recent paradigms shift in drug discovery from the target-directed approach to the system biology centered strategy. A major technique rapidly growing in interest with respect to the detection of cellular interactions is surface plasmon response (SPR). This method was introduced in Chapter 1 and is summarized briefly here. The technology exploits evanescent waves to characterize molecular or cellular interactions on the sensor surface. SPR instruments incorporate a prism to direct a wedge of polarized light, covering a range of incident angles, into a planar glass substrate covered with an electrically conducting metallic film to excite surface plasmons. The resulting evanescent wave interacts with, and is absorbed by, the electron clouds in the metallic layer, generating electric charge density waves (the surface plasmons), and thus causing a reduction in the intensity of the reflected light. The resonance angle at which this intensity minimum occurs is a function of the refractive index of any biological moiety present at the sensor surface (Figure 5.9(a)). As outlined in Chapter 1, interactions at the sensor surface cause shifts in the resonance angle, which is depicted again in Figure 5.9(b).

In applications involving the label-free measurements of neural electrical activity, the method, unlike electric detection, is artifact-free. Electrical recording involves stimulation, which generates artifacts in the detecting signal. Furthermore, voltage-sensitive fluorescent dyes are expensive, toxic, involve time-consuming labeling procedures and are affected by photobleaching. From an optical sensing point of view, changes in scattering and birefringence of a nerve can be correlated with neural activity. When an action potential propagates through an axon in the nerve, reorientation of molecular dipoles across the membrane occurs altering the refractive index of the axon membrane. In addition, action-potential propagation produces an osmolality difference across the membrane, which in turn leads to cellular swelling, an increase in the cell volume through the influx of water molecules. These alterations in the refractive index and microanatomy of the nerve result in optical scattering changes. It should be noted that the birefringence change

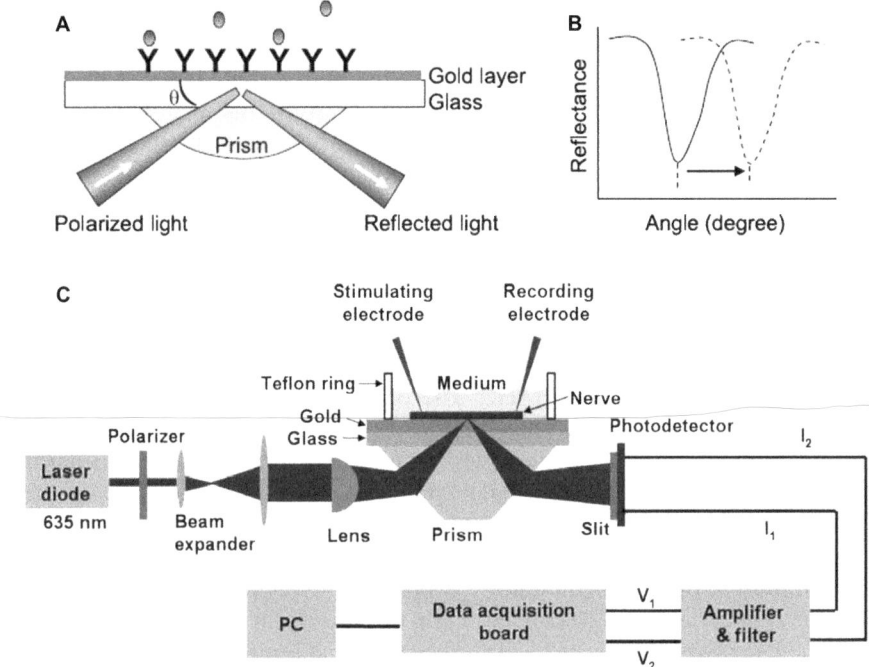

Figure 5.9 Operational principles of SPR system employed or biochemical detection and measurement of neural activity. (A) Target molecules bind to the specific, surface immobilized receptors. (B) The specific interaction is detected as a shift in resonance angle. (C) Schematic of the SPR system used to measure neural activity.[20]
(Reprinted by kind permission of the Optical Society of America.)

associated with nerve activation presumably comes not only from conformational changes of macromolecules, but also from cellular swelling described above.

The experiment presented in Figure 5.9(c) shows simultaneous electrical and optical detection at the metal–nerve interface. Electromagnetic waves propagate through the conductor–dielectric interface. Detection is based on the highly sensitive resonance of the SPR sensor and takes place in the very small volume represented by the thin layer on the surface. The experimental setup was based on the attenuated total reflection configuration, in which an incident beam was coupled through a BK7 prism on a glass slide. An SPR sensor chip with a 50 nm thick gold-coated film on a microscopic BK7 glass slide was modified for use as a recording chamber using a Teflon ring. A low-noise laser diode with a wavelength of 635 nm and an output power of 5 mW was used. The laser beam was focused onto a point with a beam diameter of 100 μm. An intensity-based SPR sensor was used because its sensitivity is reported to be equal or superior to that of a phase-sensitive method. The reflection intensity of the laser beam was detected through a multichannel photodetector array. The output currents (I_1, I_2 in Figure 5.9(c)) from two vertical elements of the photodetector array were converted to voltage signals (V_1, V_2), and amplified with a gain of 10 000 before low-pass filtering at the cutoff frequency. fc, of 500 kHz. The SPR system was initially aligned such that those two signals were equal in magnitude, and then the difference between the two signals was monitored to remove common mode noise. With additional noise reduction, the ambient noise was less than 10 μV_{RMS} in magnitude. As mentioned above, the system was built to record both optical and electrical signals simultaneously in response to electrical stimulation. A Teflon-insulated Pt–Ir wire electrode was placed on the proximal end of the nerve, while another wire was located on the distal end of it. The laser beam for the SPR detection was located between the two wires. Electrical responses were amplified a thousand times using a differential AC amplifier, filtered between 1 Hz and 10 kHz, with a 60 Hz notch filter and digitized at 20 kHz. The optical and electrical signals were simultaneously recorded and time labeled by a single DAQ board. For neural recording experiments, sciatic nerves from knee (distal) to spinal cord (proximal) were dissected from male Sprague–Dawley rats. Changes in the refractive index unit were detected when the nerve was stimulated with ethanol or nerve-blockers were applied (lidocaine 2%).[20]

An additional optical methodology is provided by light-addressable potentiometric sensor technology (LAPS).[21] In this case cells are anchored on a semiconductor surface and the local surface potential is measured when a small area on the surface is illuminated by a laser beam. The system is depicted in Figure 5.10.

Detection takes place in the electrolyte–insulator–semiconductor structure. The illumination of the silicon substrate leads to energy band transitions and the generation of electron–hole pairs. The modulation of light induces an AC photocurrent, which depends on the surface potential. When drugs are pumped in, the cells react—for example, in neurons, if a given stimulus generates a depolarization greater than the threshold, an action potential is induced, which in turn changes the bias voltage and can be detected by the sensor. Scanning or multi-light LAPS microphysiometers based on this effect can monitor cellular acidification or Na^+, K^+, Ca^{2+} and Mg^{2+} changes in concentration, or the effect of applied drugs.

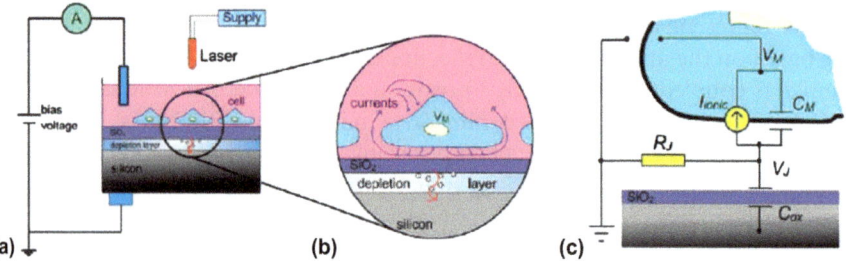

Figure 5.10 The principle of LAPS. (a) Schematic of the cell-based biosensor. (b) The cell-semiconductor interface. (c) Equivalent circuit model.[21] (Reprinted by kind permission of Elsevier BV.)

The LAPS system can be used in monitoring the K^+ induced preconditioning in cultured neurons to assess the role of N-methyl-D-aspartate (NMDA) and alpha-7-nicotinic acetylcholine receptors.

5.2.5 Acoustic Wave Detection

The thickness shear mode (TSM) device was introduced in Chapter 1. Briefly, a resonant shear wave is produced in a piezoelectric material such as quartz by electrodes (usually gold) placed on each face of the sensor. When operated in a liquid environment, a portion of the acoustic energy is not reflected at the interface but is dissipated into the liquid. The propagation of the acoustic shear into the liquid and resultant energy dissipation is detected *via* variations in the motional resistance (Rm) and series resonant frequency (fs) parameters of an impedance analyzer attached to the system. A typical experiment involving study of neurons to the gold electrode of a 9 MHz TSM is depicted in Figure 5.11. The sensor system is incorporated into a temperature-controlled static or flow-through confutation which allows the introduction of various reagents in order to detect response of the cells. The latter are kept in a CO_2 environment.

One of the first TSM studies of neuron behavior involved immortalized hypothalamic mouse cells (N-38) in terms of their response to various known neurotrophic factors such as forskolin, betaferon and cerebrolysin.[22] Betaferon has been used in the treatment of relapsing forms of multiple sclerosis to reduce the frequency of exacerbations. It is believed that the molecule acts by interacting with specific receptors found on the surface of cells. The binding of betaferon to these receptors induces the expression of a number of gene products and markers, including 29,59-oligoadenylate synthetase, protein kinase, beta 2-microglobulins, neopterins, and indoleamine 2,3-dioxygenase.

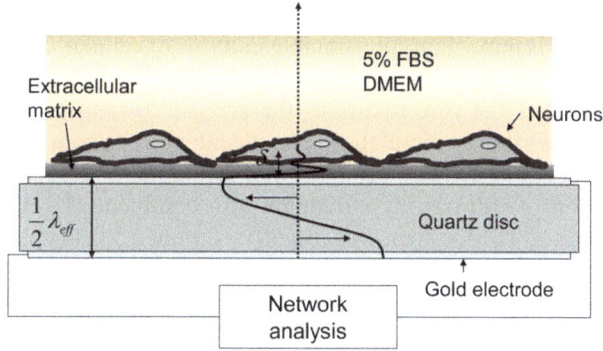

Figure 5.11 Schematic of a TSM system for the study of neurons. Cells are constantly bathed in a fresh supply of Dulbecco's modified eagle medium (DMEM) supplemented with 5% fetal bovine serum (FBS). All test solutions being made up using either 5% FBS DMEM or DMEM as the solvent to prevent adverse effects to cells.[22]
(Reprinted by kind permission of the Royal Society of Chemistry.)

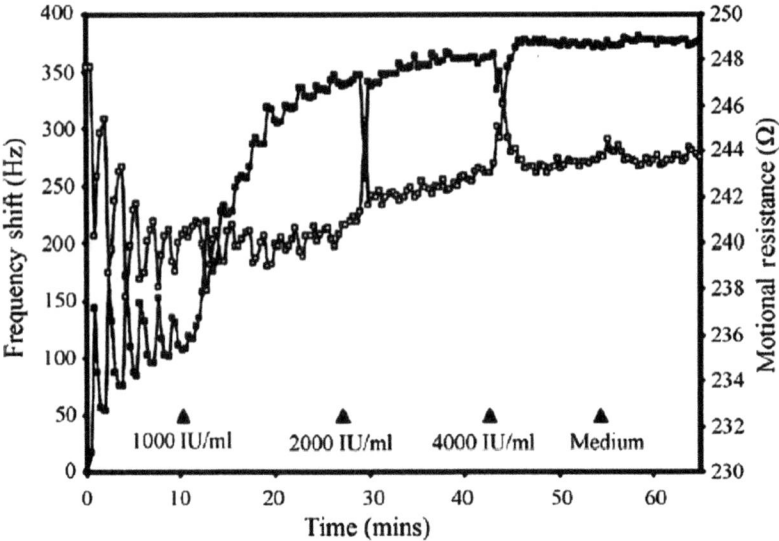

Figure 5.12   Effect of neurotrophic betaferon on a neuron population immobilized on a TSM surface. Synchronous oscillations are detected both on the resonant frequency (dark squares) and motional resistance (open squares).[22]
(Reprinted by kind permission of the Royal Society of Chemistry.)

The acoustic response of the TSM-attached neurons to betaferon is shown in Figure 5.12. Interestingly, the resonant frequency decreases considerably with little effect being observed for the response in resistance. Furthermore, the two-minute period oscillations apparent before introduction of the drug are damped as a result of interaction with betaferon. (Such oscillations are discussed in more detail in the next section.) Several questions are posed by these observations. First, experimental work has shown that the confluence of the cells on the device surface is important, presumable in terms of cellular communication. Second it is not clear how the molecule is affecting the behavior of the cells and, indeed, what is the true nature of the acoustic response to this process. This preliminary work has stimulated a significant effort to connect the acoustic response of neurons to cellular behavior at a fundamental level.

Depolarization of cultured hypothalami, at a confluence level in the range 80–100%, using the TSM system was recently examined.[23] This phenomenon is dependent on voltage-gated Na^+ and delayed K^+ channels. Normally, a stimulus that causes sufficient depolarization causes voltage-gated Na^+ channels to open and cause further depolarization (self-amplifying process) until the Na^+ equilibrium potential is reached. At this point, two mechanisms act in concert, namely the inactivation of the Na^+ channels and the opening of the K^+ voltage-gated channels to restore the cell to its resting potential. Stimulation by the addition of KCl resulting in cellular depolarization has long been examined by conventional patch-clamp techniques. It is suggested that an addition of KCl affects the electrochemical gradient of K^+, resulting in a slower

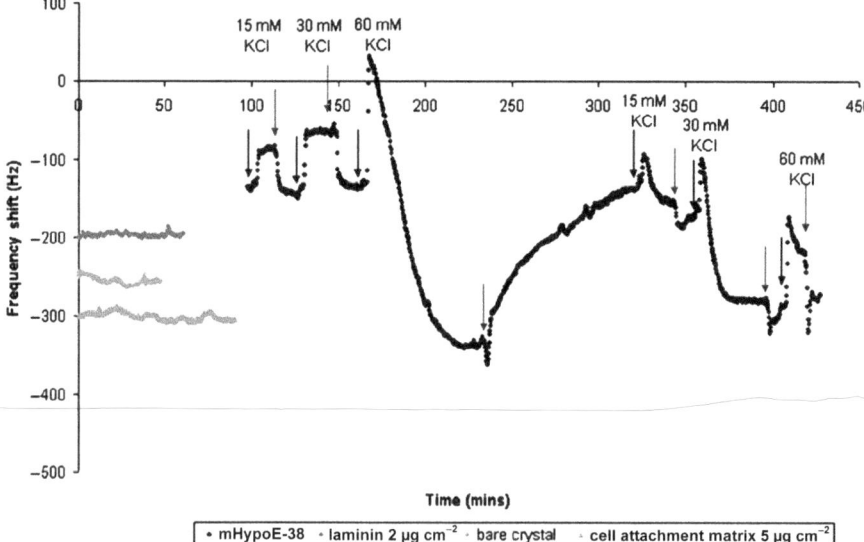

Figure 5.13 Series resonant frequency shifts for neurons held at 37 °C with addition of KCl. Black arrows indicate the addition of 15 mM, 30 mM, 60 mM and repeat sequentially. The gray arrows indicate wash-offs. Controls are shown to the left, from top to bottom: laminin 2 µg cm^{-2}, bare crystal and cell attachment matrix 5 µg cm^{-2}.[23]
(Reprinted by kind permission of the Royal Society of Chemistry.)

efflux of K$^+$ and depolarization of the membrane. Figure 5.13 shows the frequency response of the cultured neurons to additions of KCl. There are clear shifts in both fs and Rm. Typical changes in fs for 15, 30 and 60 mM KCl additions were 54, 80 and 142 Hz, respectively.[23] Analogous changes in Rm for these KCl additions were 7, 13 and 23 ohms, respectively.

As for the experiments described above, the precise relationship between the acoustic physics and the depolarization event begs an explanation. In this regard it is appropriate to consider the nature of propagation of the acoustic at the cell–sensor interface. The penetration depth of the acoustic wave (decay length) is of the order of 500 nm in the presence of cells (for a cytoplasmic viscosity of 5 cP, five times that of water). Accordingly, the acoustic wave will probe the extracellular membrane (ECM), the cell membrane (lipid bilayer) typically 10 nm, and also a part of the cytoplasm of the cells. The results, therefore, appear to indicate that the sensor is detecting not only changes in charge characteristics but also cellular structural alterations. Possibilities for the latter may well result from perturbation of coupling between the ECM and cell, together with membrane viscoelastic effects.

Research has also been conducted on the synchronization of the circadian rhythm generator of neurons with reference to the effects of glucagon.[24] Under the influence of glucagon, the neurons (two lines, mHypoE-38s and mHypoE-46) display both short- and long-term changes The effect of synchronizing the neurons prior to glucagon stimulation did not influence the cellular changes observed. The process of partial and full synchronization of the cells resulted in different responses. For full synchronization, the addition of the serum bolus triggered resonant frequency and motional resistance shifts of +75 Hz and +18.5 ohms, respectively, which decayed back to baseline levels after 30 mins. The duration of this decay closely matched the time required for full synchronization in a separate study.[24] The changes observed for partial synchronization were significantly different from full synchronization as the baseline levels in both resonant frequency and motional resistance were not re-achieved indicative of the cell–sensor system detecting the difference between full and partial synchronization. Immunocytochemistry and reverse transcriptase–polymerase chain reaction (RT-PCR) studies on these cells supported the results obtained with the TSM.

Finally, with respect to the experimental use of acoustic wave physics, the interfacial behavior of hypothalami has been investigated.[25] The surface attachment of the two cell lines outlined immediately above were investigated in terms of solution flow, the absence of serum proteins, the effect of reducing specific cell–surface interactions and the disruption of the neuronal cytoskeleton components. With respect to adhesion and deposition of neurons, f_s and Rm shifts were clearly correlated to the amount of adhered neurons on the sensor surface, whereas non-adhered neurons did not produce any significant change in the monitored parameters. In the absence of serum proteins, initial cell adhesion was followed by subsequent cell death and removal from the sensor surface. The presence of the peptide, GRGDS, was observed to significantly reduce cell–surface specific interactions compared with the control of SDGRG and this produced fs and Rm responses opposite in direction to that observable for cell adhesion. Cytoskeletal studies, using the drugs nocodazole (10 mM), colchicine (1 mM), cytochalasin B (10 mM) and cytochalasin D (2 mM) all elicit neuronal responses that were by phalloidin actin-filament staining.

5.2.6 Origin of Oscillations and Neuronal Resonance

We choose here to comment on neural oscillations in the light of those described above that are detected by acoustic physics. The mechanism of oscillations and neural resonance is fundamental for the functioning of the human brain. Berger was the first to record the brain rhythms in 1929.[31] Since then, electroencephalographic patterns have been studied intensively during different states of consciousness, such as wake and sleep intervals, hypnosis, anesthesia or epileptic states, when loss of consciousness is expected. Later, the interest in associating brain oscillations to complex cognitive operations diminished significantly. Recent observations on single neurons capable of oscillating and resonate at specific frequencies within neural networks have renewed the interest in oscillatory explanations of cognition.

Complex analysis shows that oscillatory patterns during sleep can be related to those occurring during a previous wake period, although at smaller amplitudes. This introduces the possibility that perception and temporal representation might be the result of such synchronized network activity that spans over five ranges of magnitude in frequency. Oscillations and oscillatory synchronicity are the most energy-efficient mechanisms through which information is conveyed. The non-dissipating solitary wave is an example of such a mechanism. Acoustic–electromagnetic (optical, conformational, rotational, oscillating) solutions are the fundamental components of quantum wave matrices in living molecules. How the brain stores patterns and generates creativity using so little energy might be explained by such an energy-effective mechanism.

Brain oscillations appear to be the result of a finely tuned and optimized interplay between the intense intracellular activity and the dynamic properties of neural networks and circuits. Neurons interconnected in neuronal networks and circuits behave surprisingly like miniature electrical oscillating circuits.[26] Resonance, oscillation and the intrinsic frequency preferences of neurons are due to their intrinsic resistivity, membrane capacity and the conductance of ionic channels which all decide the resonant behavior. Figure 5.14 shows an analogy between classical electronic components forming an electric resonator and the neuron. An impedance amplitude profile using input complex signals containing all the frequencies of interest shows, by fast Fourier transform (FFT), that the neuron, like a miniature electronic circuit, can discriminate between input frequencies and intrinsically select preferred frequencies.

Rhythmic oscillations are a basic feature of the membrane potentials found in spontaneously active neuronal cells or neuron networks. These cells or networks generate the patterns responsible for walking, breathing, chewing and other rhythmic movements. There are several mechanisms through which oscillatory activity can be produced. These include interactions among ion channels, inhibitory interactions among neurons in cyclic networks, cascades of metabolic reactions and/or cyclic transcriptions of genes. These mechanisms each operate with different time periods. For example, one such mechanism is established by the action of the hyperpolarization-activated current, a cationic current critical for the neuronal pacemaker activity. It is a slowly developing inward current (depolarizing) activated by hyperpolarization of the membrane beyond the resting potential and produced by a mixed Na^+/K^+ conductance.

The time constant of current activation varies from 1–2 s at close to rest potential to 100–400 ms at maximum hyperpolarization. An increase in intracellular Ca^{2+} regulates the current such that it operates at more depolarized membrane potentials.[27]

Another example of the molecular underpinnings of oscillatory behavior is related to the action of a specific glutamatergic receptor. In pacemaker neurons, the NMDA glutamate receptor generates the pacemaker rhythms and provides a mechanism for synaptic plasticity. NMDA is an amino acid derivative that acts as a specific agonist to the NMDA receptor and therefore mimics the action of glutamate on that receptor. In contrast to glutamate, NMDA binds to and regulates the above receptor only and does not bind to any other glutamate

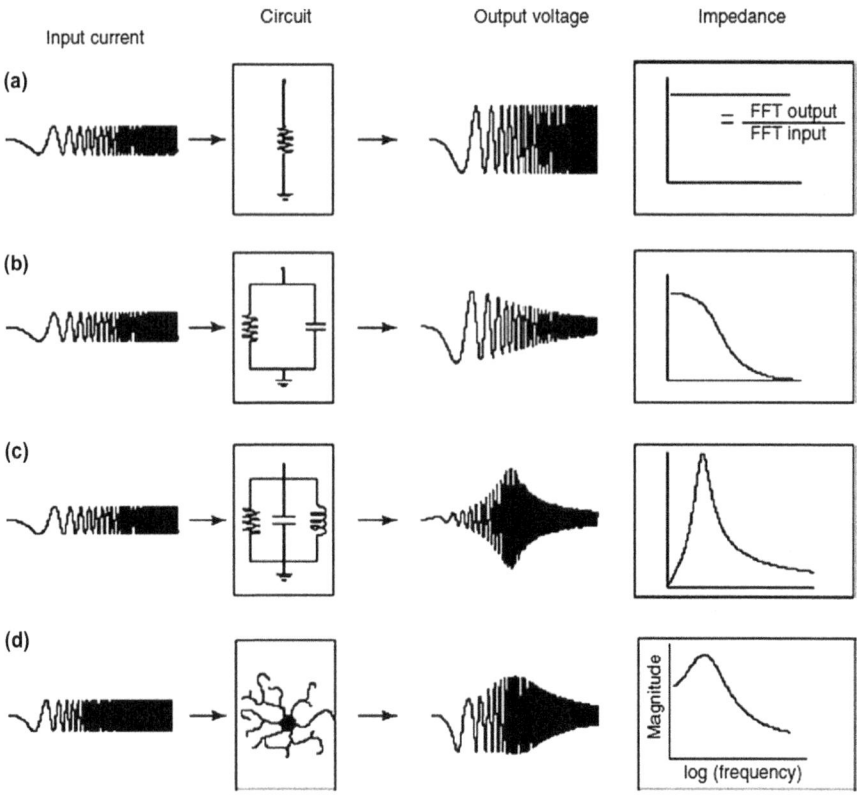

Figure 5.14 Illustration of selective properties of a resonant neuron (d) in comparison with the frequency response of electronic components: (a) resistor; (b) resistor in parallel with a capacitor and an oscillatory circuit formed by a resistor; and (c) a capacitor and an inductance in parallel. The input is a complex signal containing a range of frequencies, while the output is analyzed using the Fast Fourier Method.[30]
(Reprinted by kind permission of Elsevier BV.)

receptors. NMDA is a water-soluble synthetic substance that is not normally found in biological tissue. First synthesized in the 1960s, NMDA is an excitocin which has applications in behavioral neuroscience research. Researchers apply NMDA to specific regions of an animal brain or spinal cord and subsequently test for aberrant behavior. If behavior is compromised, it suggests that the destroyed tissue was part of a brain region that made an important contribution to the normal expression of that particular behavior. The cation pore of the NMDA receptor allows the diffusion of both Ca^{2+} and Na^+, although these processes are blocked by the presence of Mg^{2+}. When the membrane is strongly depolarized, Mg^{2+} is cleared away and Ca^{2+} and Na^+ are allowed to diffuse through the membrane. As a result of the Ca^{2+} influx, intracellular Ca^{2+} or Mg^{2+} activates a Ca^{2+}-dependent K^+ current that eventually hyperpolarizes the membrane when there is a sufficient build-up of the K^+ efflux. Membrane repolarization closes the Ca^{2+} channel. The K^+ current gradually decays as Ca^{2+} is pumped out of the cytoplasm. Provided that the neuron is receiving tonic excitatory input from some glutamatergic source, the NMDA receptor will re-open after the hyperpolarization has dissipated. The NMDA receptor transforms a steady excitatory drive into an alternating rhythm of bursts of spikes separated by silent periods of hyperpolarization, a very useful feature of pattern generators. The rhythm of NMDA oscillators is in the range of 0.1–3 Hz.[28,29]

Simple pacemaker neurons cannot produce complex rhythmic patterns of activity. Synchronized oscillations can be created either by neurons sharing or distributing the timing functions among themselves through mutually exciting or inhibiting one another *via* synaptic connections. When large numbers of neurons are involved and phase relationship requisites are met, the cyclic behavior is best produced by an interactive network of neurons, the network oscillators. Breathing, locomotion and other rhythms are generated by reticular networks that allow for changes in relative timing and in phase magnitude with respect to the other constituent neurons of that particular network. The basic design of an oscillatory network involves a diffuse excitatory input from an extraneous source so that the discharge is automatic when neurons are not being inhibited. Such inhibitory connections among the network of neurons determine the order and timing of the activity patterns.

Oscillating networks generally follow a ring structure with directed excitatory and inhibitory interconnections. Each node or group of neurons functions as a burst generator that drives a particular phase of the movement cycle and tends to inhibit its immediate predecessors. It is the retrograde inhibitions that are the most important and constant feature of the ring. A ring cycles freely in the direction counter to the inhibitory connections. Because the entire network is subjected to diffuse excitation, firing in any node is mainly due to the spontaneous depolarization following the removal of the inhibition. The rate of cycling is therefore dependent on the strength and duration of the inhibitory effects. The stronger and longer lasting the inhibition, the slower is the cycle frequency. The coupling of individual oscillators can occur in several ways. Phase coupling entrains the start of a cycle in one oscillator to a specific phase of another so that a travelling wave of activity will pass rhythmically. Relaxation coupling is the manner in which a group of oscillators become abruptly synchronized. Their individual cycles are concurrently reset by the simultaneous inhibition of all the individual oscillators. When the inhibition subsides, all of the oscillators resume their synchronous discharge. This is the mechanism responsible for synchronizing populations of thalamic neurons that project to the cerebral cortex. Synaptic connections between excitatory and inhibitory thalamic neurons force each individual neuron to conform to the rhythm of the group. These coordinated rhythms are then passed to the cortex by the thalamocortical axons, which excite cortical neurons. This is how a small group of centralized thalamic cells can impose synchronicity to a much larger group of cortical cells.

Membrane potential oscillations are always entrained throughout populations that are joined by gap junctions. Such oscillations occur in many neurons at sub-threshold levels, meaning that action potentials need not be triggered to maintain the oscillation. The origin of this background oscillation implies a fast persistent Na^+ current and a slow non-inactivating K^+ current. The kinetics of the latter establishes the frequency of the oscillation. The Na^+ channel is a different one than that involved in action potentials. It opens at sub-threshold potentials (-60 to -65 mV) and closes only when the membrane hyperpolarizes as a result of the ensuing depolarization-activated K^+ current. The background level of membrane depolarization determines the strength of the Na^+ current and the frequency of the oscillation. The increasing depolarization increases the Na^+ current which, in turn, increases the frequency up to a maximum of approximately 20 Hz. Synchronized neuronal assemblies in the cerebral cortex generate an oscillating membrane potential that may be part of the basis of electroencephalography (EEG) rhythms. Ion channel oscillations, membrane capacitance and resistance determine resonance behavior at a preferred frequency. The resonance of cortical neurons ensures that they respond to inputs arriving at the preferred frequency and without any time delays. They are effectively tuned to respond to specific dynamic rhythms. Such models as the ones described above might explain the oscillatory and resonant behavior of neurons, and possibly the effects of melotherapy and an individual's musical preference. However, all the molecular dynamics that produce resonance phenomena are still unknown.

This raises additional interesting questions concerning the possible relation of certain brain rhythm frequencies to particular behaviors. For example, magnetic transcranial stimulation experiments reveal that frequencies less than 1 Hz induce brain inhibition, whereas over 1 Hz they induce excitation. Brain activity at 1 Hz frequency generally corresponds to deep meditation and hypnotic phases, both of which are the only non-destructive methods for mind relaxation known today. Relaxation takes place by coordination and deep synchronization of different brain areas. A possible hypothesis is that these synchronous oscillations are somehow hardwired in the structure of cells, and can consequently be manifested in brain rhythms and thus determine human behavior.

Until now it was supposed that certain cortical centers are coordinating and entraining the oscillation of different neural networks inside the brain. It has not been considered previously that some embryonic neural cells deposited in a Petri dish are growing, extending dendrites and organizing themselves in an incredibly coherent oscillating network. This process cannot be detected visually with microscopes, but can be detected with instruments sensitive to such elusive oscillations. Short- and long-term consolidation of information and consciousness might be the result of such synchronized network activity. The different classes of oscillations and the behavior patterns related to them are not found exclusively in the human brain, but they are phylogenetically preserved throughout mammalian evolution, which would support the assumption that there is a universal mechanism at work in the brain, small or large, and the possibility that animals are, in fact, intelligent. The alpha brain rhythm (8–12 Hz) is also superimposed on the Earth's background base frequency of 7–10 Hz, suggesting that our state of wakefulness and awareness is related to the rhythms of the planet we live on. Interestingly, the gamma brain rhythm (higher than 35 Hz) can be measured using an electroencephalograph on pilots when maximally concentrating on landing a plane, but also in the brains of Buddhist monks who meditate for decades. What is this universal oscillation at specific frequencies in our brains and in the brain of superior animals?

5.2.7 The Scanning Kelvin Nanoprobe

Conventionally, a Kelvin probe of the macroscopic kind is used to detect work function levels in metals or semiconductors—the work function being the energy necessary to extract an electron from the Fermi level of the material to infinity. A much greater resolution, high-sensitivity version of the instrument, termed the scanning Kelvin nanoprobe (SKN), was developed at the University of Toronto with applications in the field of biophysics in mind.[30] The instrument is capable of imaging both the work function and the topographical maps for a scanned surface at an extremely high nanometric resolution. The evidence for the high-sensitivity performance of the instrument came from its ability to detect molecular events such as a single base mismatch in DNA. Moreover, it is a non-contact and non-destructive technique that uses low-intensity electric fields and small amplitude oscillations. As a result, this technique causes minimal perturbations to the system under observation: unlike patch-clamp techniques, it is an electrode-less method, does not require the insertion of probing electrodes into the system under study.

The instrument functions by vibrating a tungsten probe over a neuron or neuron network grown on a metallic surface connected to the conducting table which is capable of movement in the X and Y directions (Figure 5.15). The vibrating probe approaching the neurons vertically, generates a time variant capacitance resulting in the generation of a Kelvin current, shown as $i(t)$ in Figure 5.16.

By applying a null current condition, the capacitance voltage can be determined and the work function of cells can be extracted. In order to examine the potential of the SKN for the study of neurons, immortalized N-38

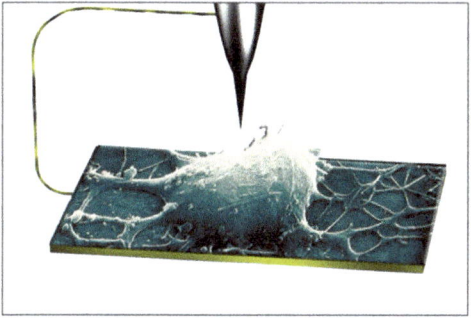

Figure 5.15 Non-invasive investigation of neuron response to drugs by scanning Kelvin nanoprobe technique.[30]

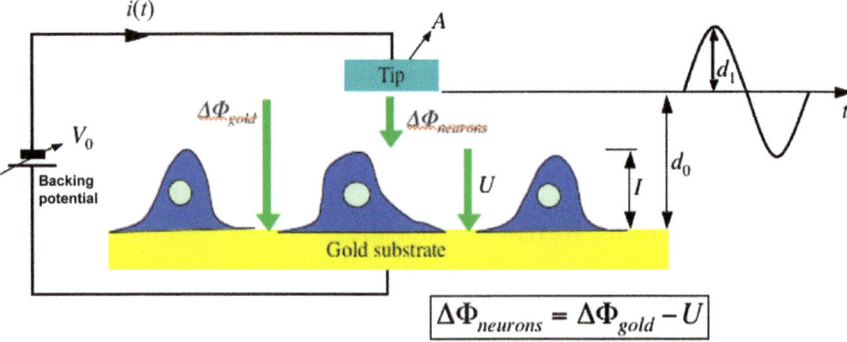

Figure 5.16 Schematic representation of the SKN detection principle for a neuron cell culture. The probe vibrates at a distance d_0, with an amplitude d_1 over the cells which are grown on a gold substrate. The presence of the neurons alters the gold work function by a potential difference reflecting the dielectric characteristics of the cell membrane and cytoplasm. The detected Kelvin current, $i(t)$, is very sensitive to any change in the metabolic state of the neurons. A is the area of the probe, V_0 is the external backing variable voltage, U is the voltage difference that reflects the dielectric properties of cell layer and the contribution of the image force potential, d_1 is the amplitude of the vibration and d_0 is the rest position of the probe; $\Delta\Phi$ is the work function and I is the thickness of neuron layer.[22]
(Reprinted by kind permission of the Royal Society of Chemistry.)

hypothalamic mouse neurons were deposited on the gold electrode of a TSM device much as described above in the section on acoustic wave sensors. For this experiment, substrates were prepared using gold-covered silicon wafers upon which a protein matrix of laminin and fibronectin in DMEM were used to provide an appropriate surface for cellular adherence.[22] Immortalized hypothalamic N-38 undifferentiated neurons were first separated from the main frozen culture and grown in a flask in DMEM supplemented with 5% fetal bovine serum (GIBCO). These were then left in an incubation chamber for 4 h at 37 °C under 5% CO_2 for cells to deposit and attach by gravity. After this time, a complete medium change was required to remove DMSO from the cell culture. DMSO was present in the initial cell culture to enable faster freezing and thawing during storage. When a substantial population was obtained, the cells in the culture medium were added to the gold surfaces separated in plastic Petri dishes. Usually the cell culture achieves 90–100% surface coverage in 48 h and is fully differentiated. There is an optimum surface coverage related to dendritic connection formation and to the fact that neurons prefer a less crowded surface in cultures.

SKN measurements can be performed in both scanning and in localized, real-time measurement modes. Figure 5.17(a) confirms the stability of the scanned signal obtained from the bare gold surface in liquid.[22] When there are neurons on the surface, the scan (Figure 5.17(b)) shows a dramatic variation in the potential ($\sim$ 550 mV). This change can be ascribed to the complex distribution of charges and dielectric dipoles on the surface of the cell. If the cells are

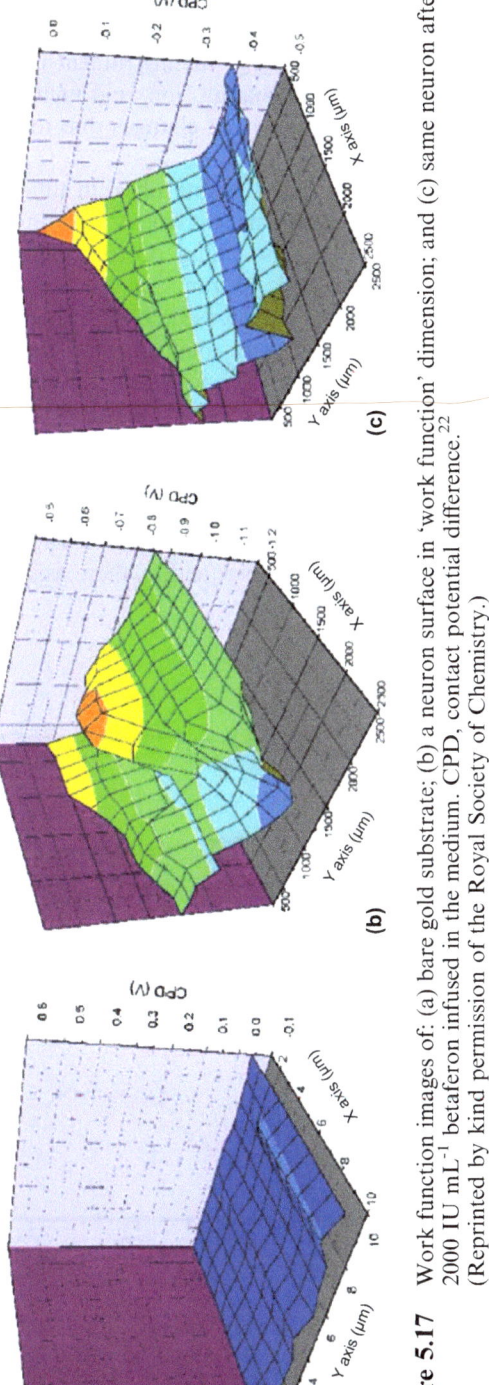

Figure 5.17 Work function images of: (a) bare gold substrate; (b) a neuron surface in 'work function' dimension; and (c) same neuron after 2000 IU mL^{-1} betaferon infused in the medium. CPD, contact potential difference.[22] (Reprinted by kind permission of the Royal Society of Chemistry.)

exposed to betaferon at 2000 international units (IU) mL^{-1}, there is a change in both the surface potential value ($\sim$350 mV), and in the overall shape of the scan as shown in Figure 5.17(c). Clearly, the instrument is detecting not only changes in the local distribution of charges, ion, and membrane permeability caused by the drug, but also subtle shape changes due to movements in the cytoskeleton matrix.

In order to focus on the alterations taking place in time with respect to a specific neuron, a real-time localized measurement at a single point over one single cell can be performed. The scanning protocol was altered to allow vibration of the probe over one single neuron.[22] In this experiment, the neuron was exposed to the same drug but at increased concentration. Time-dependent changes in the measured potential were found as shown in Figure 5.18. An abrupt 350 mV increase in the contact potential difference (CPD) can be noticed when the neuron was exposed to 1000 IU mL^{-1} betaferon, confirmed by a decrease of about 150 mV. There was no such decrease when the neuron was subjected to 2000 IU mL^{-1} of the drug; the first 100 s showed a 280 mV increase and then a slight saturation of cell receptors after approximately 500 s. For 4000 IU mL^{-1} there was a 600 mV initial decrease of the signal, followed by a rise to a maximum of 300 mV. Subsequently, there was a 350 mV decrease at the end of the observation period.

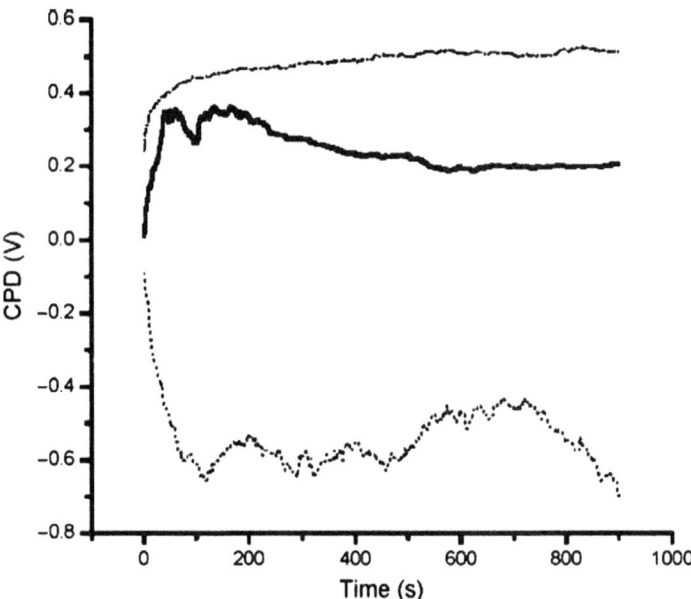

Figure 5.18 Changes in the measured work function of one particular neuron after the addition of betaferon at different concentrations. The solid line (—) is betaferon at 1000 IU mL^{-1}, the dash-dotted line (-.-) is betaferon at 2000 IU mL^{-1} and the dotted line (...) is betaferon at 4000 IU mL^{-1}.[22] (Reprinted by kind permission of the Royal Society of Chemistry.)

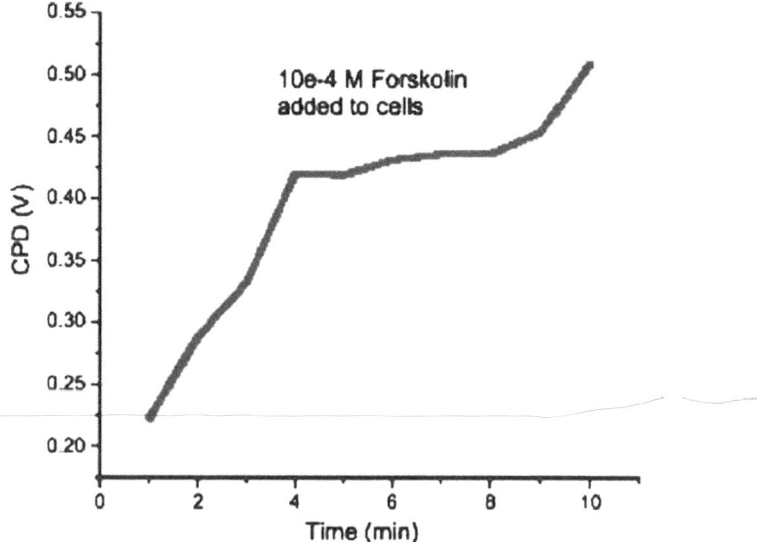

Figure 5.19 Work function variation on the surface potential of a neuron when forskolin is added to the medium.[22]
(Reprinted by kind permission of the Royal Society of Chemistry.)

It is particularly interesting to note that these SKN results correlate with the results obtained from acoustic wave detection outlined above. To further test the real-time performance of the SKN in response to drugs, forskolin was added at concentrations found to be biologically active in the acoustic wave experiments.[22] Remarkably, both SKN and TSM results correlate with a classical enzyme-linked immunosorbent assay (ELISA) analysis for the forskolin experiment.[22] The addition of 10^{-4} M forskolin produces a large 300 mV change in the surface potential of the neuron, as shown in Figure 4.19, in precise correspondence with the acoustic and classic ELISA assays. From such results, it is evident that the SKN is able to detect not only the presence of neurons, but also the changes after the addition of specific drugs. It remains to be established, however, what relationship exists between the observed alterations in CPD and neuronal response to the drugs studied in these experiments.

5.3 Future Possibilities in Cellular and Neuronal Detection

Changes in the electrical and optical properties of cells, and of course neurons, such as impedance, intensity of current, surface potential, refractive index, interfacial evanescent wave propagation and work function can be measured with very high accuracy using vibrational fields of different frequencies that are intrinsically higher than conventional biological techniques. Analytical methods based on vibrational fields can successfully complement the classical biological methods, through the focusing of a battery of different techniques on the same biological target, for a more integral characterization of the biological systems. Neural resonance and oscillations in the brain are very interesting phenomena and acoustic and electromagnetic vibrational fields of high resolution appear to be an ideal method for their study. The neuron is a naturally oscillatory circuit exhibiting resonant behavior due to the capacity and resistivity of the membrane and the conductivities of the ionic channels which determine its frequency-dependent properties.

The exact mechanism of neural oscillation and resonance is still unclear. It is certainly related to brain rhythms and their associated behavior. Biology is based in part on the movement of electrons, which are quantum entities that behave far more strangely that it is currently assumed. For example, contemporary calculations do not explain how adenosine triphosphate (ATP) in mitochondria moves electrons at such great speed along chains of intermediate molecules. Does this reflect quantum tunneling behavior? Is the process related to the vibratory nature of physiological sensing mechanisms? Scientists are intrigued by the vibratory nature of the universe from pulsating supernovae down to molecular and subatomic phenomena. Ultimately, everything has a quantum nature. The eye constructs color starting from vibratory frequencies of light by recognizing the characteristic frequencies at which certain molecules vibrate, thereby constructing a catalog of colors. The smell receptors discriminate over 100 000 odors when theoretically only 400 would be distinguished based on the molecular conformation of molecules. Is the mechanism for vibrational sensing based on quantum phenomena such as electron tunneling? Are detection methods based on vibrational fields capable of assisting the budding science called quantum biology? We hope that such advanced research in biology will have as side effects amazing applications such as better drugs, high efficiency solar cells and superfast quantum computers, while deciphering the quantum basis of biological processes and life itself. Life has a quantum basis which is not yet understood and quantum biology illustrates the efforts to apply the quantum physics in biology, 100 years after the physical reality revealed its fundamentally quantum nature.

References

1. L. F. Jaffe and R. Nuccitelli, *J. Cell. Biol.*, 1974, **63**, 614.
2. W. M. Kuhtreiber and L. F. Jaffe, *J. Cell. Biol.*, 1990, **110**, 1565.
3. B. Reid, R. Nuccitelli and M. Zhao, *Nat. Protoc.*, 2007, **2**, 661.
4. B. Reid, E. O. Graue-Hernandez, M. J. Mannis and M. Zhao, *Cornea*, 2011, **30**, 338.
5. I. Giaever and C. R. Keese, *Proc. Natl. Acad. Sci. U. S. A.*, 1991, **88**, 7896.
6. C. R. Keese and I. Giaever, *Exp. Cell Res.*, 1991, **195**, 528.
7. D. Malm, A. Giaever, B. Vonen, P. G. Burhol and J. Florholmen, *Acta Physiol. Scand.*, 1991, **143**, 413.
8. P. Mitra, C. R. Keese and I. Giaever, *BioTechniques*, 1991, **11**, 504.

9. J. A. Stolwijk, C. Hartmann, P. Balani, S. Albermann, C. R. Keese, I. Giaever and J. Wegener, *Biosens. Bioelectron.*, 2011, **26**, 4720.
10. C. R. Keese, J. Wegener, S. R. Walker and I. Giaever, *Proc. Natl. Acad. Sci. U. S. A.*, 2004, **101**, 1554.
11. C. R. Keese, K. Bhawe, J. Wegener and I. Giaever, *BioTechniques*, 2002, **33**, 842.
12. C. M. Lo, M. Linton, C. R. Keese and I. Giaever, *Cell Commun. Adhes.*, 2001, **8**, 139.
13. J. Wegener, C. R. Keese and I. Giaever, *Exp. Cell Res.*, 2000, **259**, 158.
14. A. B. Moy, M. Winter, A. Kamath, K. Blackwell, G. Reyes, I. Giaever, C. Keese and D. M. Shasby, *Am. J. Physiol.*, 2000, **278**, L888.
15. S. Dharia, H. E. Ayliffe and R. D. Rabbitt, *Lab Chip*, 2009, **9**, 3370.
16. E. Moore, O. Rawley, T. Wood and P. Galvin, *Sens. Actuators, B*, 2009, **b**, 187.
17. E. Moore, A. Paschero, W. Messina and E. McLoughlin, in *2011 IEEE SENSORS Proceedings*, IEEE, Piscataway, NJ, 2011, pp. 320–322.
18. A. Paschero, E. McLoughlin and E. Moore, *AIP Conf. Proc.*, 2011, **1326**, 37.
19. M. R. Sierks, G. Chatterjee, C. McGraw, S. Kasturirangan, P. Schulz and S. Prasad, *Integr. Biol.*, 2011, **3**, 1188.
20. S. Ae Kim, K. Min Byun, J. Lee, J. Hoon Kim, D. G. Albert Kim, H. Baac, M. L. Shuler and S. June Kim, *Opt. Lett.*, 2008, **33**, 914.
21. Q. J. Liu, H. Cai, Y. Xu, Y. Li, R. Li and P. Wang, *Biosens. Bioelectron.*, 2006, **22**, 318.
22. L. E. Cheran, S. L. Cheung, A. Al Chawaf, J. S. Ellis, D. D. Belsham, W. A. MacKay, D. Lovejoy and M. Thompson, *Analyst*, 2007, **132**, 242.
23. S. Cheung, L. J. Fick, D. D. Belsham and M. Thompson, *Analyst*, 2010, **135**, 289.
24. S. L. Cheung, L. J. Fick, D. D. Belsham and M. Thompson, *Analyst*, 2011, **136**, 2786.
25. S. L. Cheung, L. J. Fick, D. D. Belsham, D. A. Lovejoy and M. Thompson, *Analyst*, 2011, **136**, 4412.
26. B. Hutcheon and Y. Yarom, *Trends Neurosci.*, 2000, **23**, 216.
27. W. A. MacKay, *Trends Cogn. Sci.*, 1997, **1**, 176.
28. S. H. Snyder, *Nature*, 1993, **363**, 594.
29. E. W. Keefer, A. Gramowski and G. W. Gross, *J. Neurophysiol.*, 2001, **86**, 3030.
30. L. E. Cheran, P. Benvenuto and M. Thompson, *Chem. Soc. Rev.*, 2008, **37**, 1229.
31. H. Berger, *Arch. f. Psychiat.*, 1929, **87**, 527–570.

CHAPTER 6

The Biomimetic Interface between Brain and Electrodes: Examples in the Design of Neural Prostheses

6.1 The Nature of the Device–Brain Interface

Both neurons and electronic devices process information in the same all-or-none impulses of electricity, though action potentials and logical states, respectively. Electricity is carried as ion fluxes in the brain and as electron currents in metal conductors in electronic circuits. As we have emphasized throughout this text connecting any artificial device to cells, and relevant to this section of this text, specialized tissues of the brain, presents a major biocompatibility problem. With respect to brain implants, it is clear that extreme precision will be required in order to reach the target tissue and, additionally, maintaining the implant in place with minimal damage being inflicted on surrounding areas. Any attempt to use implantable devices to replace damaged or non-functioning nerves is met with extreme stability and biocompatibility problems associated with the tissue–device interface. However, it is possible that novel interface technologies could potentially provide new therapies for the restoration independence for severely disabled patients, a strategy offer referred to as neural regeneration. Just as important, the regeneration approach can have significant implications for neuroscience in that it facilitates a better understanding of normal and pathological aspects in the functioning of the central nervous system (CNS).

Prosthetic structures use electrical or optical stimulation to substitute for lost biological function. The development of biomimetic prostheses began in 1958 when the pacemaker was introduced. Since there is no interaction between the pacemaker and the soft tissue, other than the need for a precise electrode site, biocompatibility issues are not as critical compared with other implantable devices.[1] The cochlear implant is a successful neuroprosthetic device. It is routinely used in cases where the auditory neurons remain intact, *e.g.* when hair cells are lost and the implant directly stimulates the auditory neurons. For this application, microfabrication techniques generate electrode arrays (12–22 electrodes) using biocompatible materials with micron-size pores where ingrown tissue develops, naturally fixing the prosthesis in place. Problems with the implant arise approximately three weeks after surgery when fibrous tissue grows to a thickness of about 400 µm, consequently increasing the local electrical impedance. The impedance must be as low as possible so as not to impede the transfer of the functional signal.[2] After three weeks, the adsorption of non-specific tissue around the implant intensifies until vascular tissue with well-developed capillaries envelopes the implant surface; nevertheless, the prosthesis remains functional for a very long time.[3,4]

The interaction between electrodes and brain tissue is critical and determines the functional performance of the electrode. The electrode material should not be toxic, it should be stable and should have superior electric properties. Metal and silicon electrodes are commonly used for their electrical properties, but their mechanical properties create a high strain field at the interface, which maintains the inflammatory response of the tissue. Softer electrodes, made from polyimide for example, pose insertion difficulties since they lack the stiffness necessary to penetrate the outer layers and buckling will only extend the trauma to the tissue. Self-dissolving rigid coatings may be a solution for this problem The size of the electrodes should be as small as possible to minimize tissue damage, but the electrode impedance is inversely proportional to the surface area and sometimes a larger area of neurons has to be stimulated or sampled, so there is a design constraint regarding the dimension of the inserted electrodes.

As we have implied previously, consideration of the surface properties of devices for implantation is crucial. Surface modification using the extracellular matrix (see Chapter 2) similar to that present in neurons—fibronectin, collagen, laminin or peptide fragments from proteins such as RGD (Arg-Gly-Asp) IKVAV (Val-Ala-Val) or YIGSR (Tyr-Ile-Gly-Ser-Arg)—significantly improves cell adhesion, morphology and differentiation compared with untreated surfaces. Surface topographical features such as grooves, ridges, pillars and holes are much appreciated by the neuron cells, for reasons that are still unclear.

Implanted electrodes are usually unable to sense consistent neuronal signals for more than a few months. The loss of sensitivity is largely associated with the continuous electrochemical processes taking place at the interface. Detected or applied currents can harm living cells through such electrochemical reactions occurring at the electrode interface. These processes further lead to protein attachment and resultant tissue build-up on the electrode surface that masks the input signals from neurons, either through an increase in electrical interfacial impedance or by causing the death of the contact neuron. These persistent challenges limit the efficacy of neuronal implants.

The damage begins at the moment of insertion, when the electrode is implanted through the meninges, *via* a hole made in the dura mater, through the arachnoid and the pia matter, en route to the brain parenchim, where the vascular density is of about 160 capillaries per mm^2. Vascular damage happens immediately by tissue and fluid displacement, vessel rupture, severing and dragging, and breaking of cell bodies. A brain edema is formed locally by serum proteins and by infiltration of cells such as neutrophil granulocytes, T-lymphocytes and blood-borne macrophages, generating an inflammatory response at the site of insertion. The wound healing mechanism is subsequently initialized through thrombin release, nitric oxide (NO) production generating oxidative stress, cytokine release and activation of microglia, the first line of defense. The microglia engage in phagocytic activity similar to the role of macrophages in non-CNS tissues, engulfing and digesting the cellular debris generated by the injury, the molecules released by the dying neurons and the plasma constituents of the blood infiltrated by capillary ruptures. Astrocytes (which account for 30–65% of the glial cells) play a key role in forming a physical barrier around the inserted electrode, known as the glial scar which is, from the implant perspective, the undesirable component because it raises the interfacial impedance.

Chronic implantation renders the glial scar more compact and confined to the electrode. The continuous presence of the implant induces the formation of a sheath composed of reactive glial cells, probably due to the 'frustrated phagocytosis' process: macrophages meeting a foreign object in the body start secreting lytic enzymes after interrogating the object in order to degrade it. If the object is too big or non-degradable, the macrophages fuse into multinucleated giant cells and continue to secrete superoxide agents and free radicals, increasing inflammation and therefore exacerbating the neuronal damage. Moreover, the mechanical friction due to brain micromotion arising from physiological processes such as vascular pulsatility, cardiac rhythm and respiratory pressure, spontaneous movements of the head, *etc.* can slightly displace the electrodes—compressing, extending or even tearing the adjacent soft tissue. Astrocytes sense mechanical stress and translate it into chemical reactions. Increase in intracellular free sodium and calcium, activation of phospholipases, free radical formation and secretion of potent antigens are all causing delayed neuron depolarization and transient neuronal dysfunction. Since any exacerbation of glial scar formation impedes and compromises the recording or stimulation stability of the electrode, strategies to minimize the above responses to the invading electrode must be found. Inactivating or blocking the cellular receptors at molecular level, as well as the local delivery of antibodies, antagonists and anti-inflammatory agents to surpass the initial acute stage of the inflammation and attenuate the cellular response, or releasing neurotrophic factors to guide and sustain the nearby neurons might all constitute molecular-level solutions in an attempt to modulate and optimize the tissue-electrode interface.[5–10]

To be effective and move beyond the realm of science fiction, regenerative medicine must be built on a profound understanding of human biology both from a molecular and a holistic perspective, based on an integrated, interdisciplinary and coherent knowledge of the latest advances in nano- and microtechnologies, bioelectronics, bioenergetics, biophysics, biochemistry, neurology, chemistry, engineering, information processing, computer science and other correlated sciences.

6.2 Electrode–Tissue Interface in Deep Brain Stimulation

6.2.1 Electrode Implantation

Since the 1990s, deep brain stimulation (DBS) has become an increasingly acceptable technology. It represents the current approach for treating debilitation clinical conditions such as movement disorders related to Parkinson's disease, essential tremor, dystonia and epilepsy, all of which cannot be solved with less invasive methods. Implantable neurostimulation has been previously applied as spinal cord stimulation for pain management and hearing augmentation in cochlear implants, as discussed above. Since an unexpected positive side effect of this treatment is the improvement in the patient's mood, the method has been adapted to treat severe cases of depression where the individual does not respond to medication, psychotherapy or electroconvulsive treatment. It has also been used to treat Tourette's syndrome, phantom limb pain and obsessive–compulsive disorders, although it is not clear if the long-term benefits outweigh the risks in these particular cases. Exciting results have, however, been obtained in post-traumatic coma cases, when minimally conscious patients regained some ability to speak, move and be fed without depending on a gastrostomy tube.[11–13]

There is a limited understanding of the mechanisms by which DBS achieves such beneficial effects. In neurological conditions such as Parkinson's disease, brain function is severely impaired by synchronization processes. The resting tremor happens with a frequency dictated by neurons in the thalamus and the basal ganglia, which fire in a synchronized and periodically manner. In normal conditions these neurons fire incoherently when activating premotor areas and the motor cortex. At the present time, when patients with advanced Parkinson's no longer respond to drug therapy, electrodes are surgically implanted in target areas of the brain and electrical deep brain stimulation is applied with a frequency above 100 Hz. The technique was developed empirically and mimics the effects of tissue lesioning by suppressing neuronal firing, which in turn suppresses the peripheral tremor. DBS causes nearby astrocytes to release adenosine triphosphate (ATP), a precursor to adenosine, through a catabolic process. In turn, adenosine A1 receptor activation depresses excitatory transmission in the thalamus, this causing an inhibitory effect that imitates ablation or lesioning. In contrast with thermocoagulation, the procedure is reversible

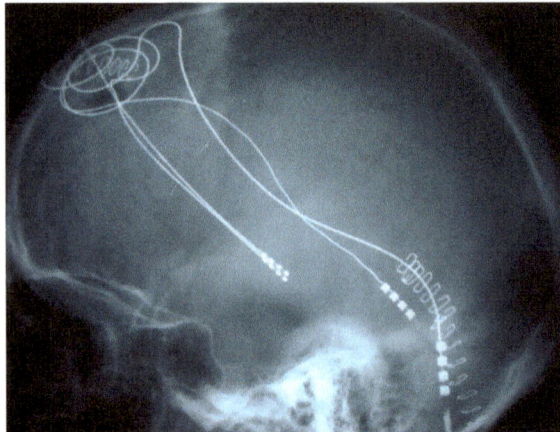

Figure 6.1 X-ray image of a deep brain stimulation electrode.
(Kindly provided by Russell J Andrews, MD, Smart Systems and Nanotechnology, NASA Ames Research Center.)

and has fewer side effects (not negligible though), such as dysarthria, dysesthesia or cerebellar ataxia.[14]

The brain pacemaker consists of a pulse generator, an electrode inserted in the brain tissue and the connecting wire that is wrapped around the vagus nerve in the neck (Figure 6.1).

The pulse generator is inserted into an opening in the chest or abdomen. The neural pacemaker is destabilized in a controlled and coordinated manner, usually with a first strong stimulus to reset the cluster of firing neurons, followed by a weaker stimulus administered with a time delay that desynchronizes the network by hitting the cluster in its most vulnerable phase. To minimize local effects on the neural tissue, a deeper understanding of the mechanisms taking place at molecular level still need to be achieved. Minimizing energy consumption by optimizing signal intensities is also essential, since the implanted pacemaker battery has a limited lifetime.

It is difficult to predict or to measure the effects of stimulation due to the absence of parallel advances in the engineering design of the electrodes and a scientific analysis of the technologies involved. It is now widely accepted that the electric field depends on the shape of the electrodes, the distribution of anodes and cathodes, and the biophysical properties of the tissue. The neural response to the electric field is related to the second derivative of the potential distributed along each neuron.[15,16] The therapeutic response to DBS is directly linked to this axonal activation. Figure 6.2 shows the electrical model of the axon, including the Ranvier nodes, the myelin attachment segments (MYSA), the paranodal main segments (FLUT) and intermodal segments (STIN), the fast (Na_f) and persistent (Na_p) sodium, slow potassium (K_s) and linear leakage (L_k) conductances in parallel with the nodal capacitance (C_n). G_m in parallel with C_m represents the myelin sheath, and G_i in parallel with C_i represents the intermodal axolemma.

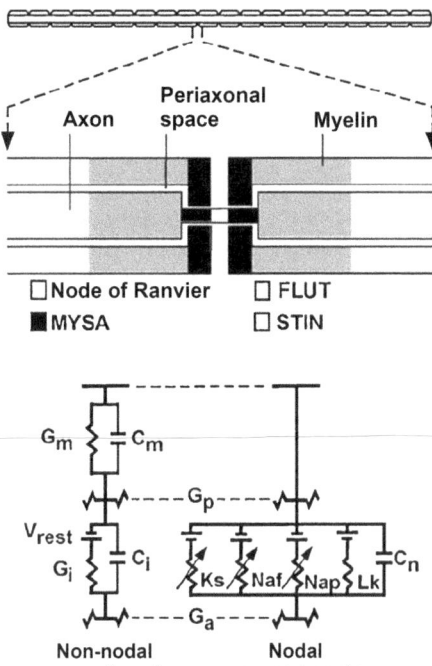

Figure 6.2 Equivalent electrical circuit for a neural axon.[16]
(Reprinted with kind permission of the International Federation of Clinical Neurophysiology.)

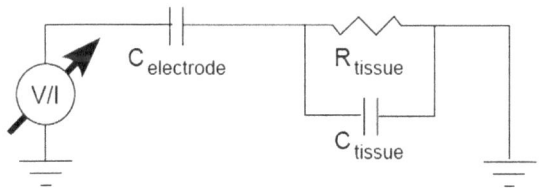

Figure 6.3 Electrical circuit model for the electrode–tissue interface.[16]
(Reprinted with kind permission of the International Federation of Clinical Neurophysiology.)

The implanted electrode is not a perfect source of voltage or current and the surrounding tissue is not a perfect conductor as it is shown in the model. Electrode effects and tissue capacitance must be carefully considered in order to avoid spatial errors as small as 1 mm for activation, which can lead to dramatic consequences for the patient. The electrical circuit model for the electrode–tissue interface is presented in Figure 6.3 where C_{dl} is the double-layer capacitance of the electrode, in parallel with the simplified tissue impedance (R_{tissue} and C_{tissue}).

Fourier finite element methods are used to study and evaluate the effects of the tissue and electrode properties on the volume of activated tissue during deep brain stimulation. The distribution of potential as a function of time and space is calculated for a range of stimulus waveforms and then applied to the model of myelinated axons to determine neural activation. Electrode geometry is also essential. Even though the different target nuclei offer considerable differences in their anatomical structure, only limited types of electrodes are currently available. Microwires, glass electrodes, polymer electrodes and silicon micromachined array of electrodes are the most popular. For example, metallic electrodes consist of four cylindrical contacts with a height of 1.5 mm and a diameter of 1.27 mm. The distance between the contacts is 0.5 or 1.5 mm. The contacts can be activated separately, and the two commonly used approaches are monopolar and bipolar stimulation. Monopolar stimulation causes a wider distribution of the current in all directions in contrast to bipolar stimulation where the current flows between the activated contacts and stimulates a more restricted area. Segmented electrodes with different active surface areas distribution as shown in Figure 6.4 have been investigated.[17]

The effects of amplitude, frequency and pulse width have been further investigated for an efficient stimulation.[18] Figure 6.5 illustrates the computer modeling of the representation of the spatial voltage of a DBS electrode used for the treatment of Parkinson's disease. The electrodes are inserted in the subthalamic nucleus. Microelectrode recordings are utilized to ensure accurate placement of the stimulating electrodes in the target tissue. The clinical changes seen with DBS in Parkinson's disease are consistently beneficial but, again, there continues to be only marginal understanding of the mechanisms by which DBS achieves these results. Analytical models of the DBS four-contact electrodes have been developed.[18] Using software developed to simulate

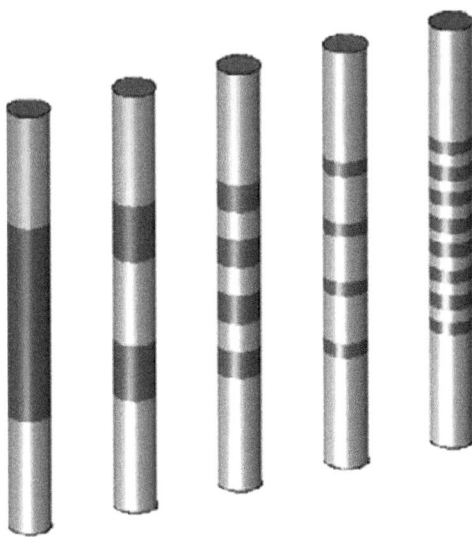

Figure 6.4 Different types of implantable segmented electrodes.[17]
(Reprinted with kind permission of the Institute of Physics Publishing.)

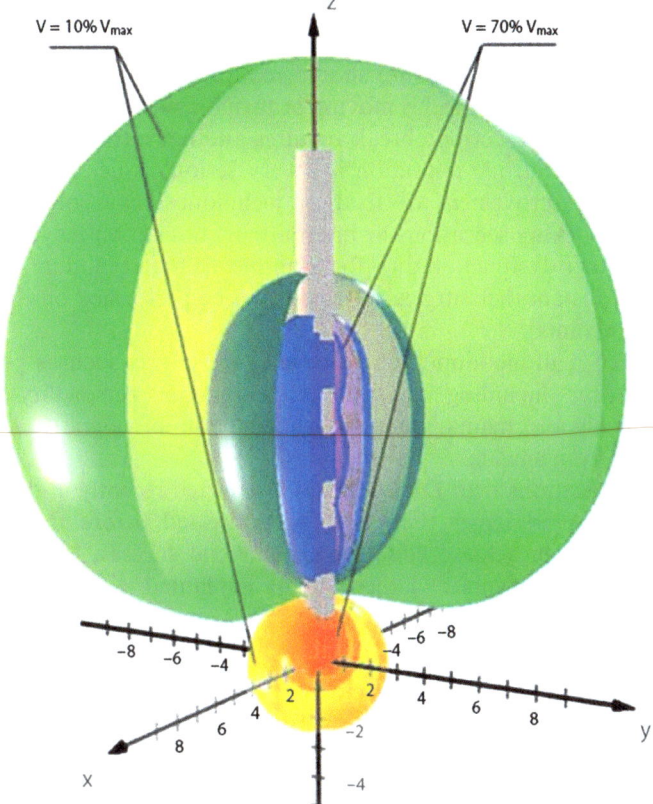

Figure 6.5 The spatial model of the field generated by a four-contact deep brain stimulation electrode.[18]
(Reprinted with kind permission of Karger Publishers.)

individual neurons and neural circuitry of the basal ganglia, the results of the model are compared with those of data obtained during DBS surgery. Firing rate, interspike intervals and regularity analyses are performed on the simulated data.

6.2.2 Device-related Complications

The greatest disadvantage of the deep brain stimulation method is the fact that risky and invasive brain surgery is required to implant the device. There is a high potential for complication during such a procedure and the risks are further perpetuated when the individual must eventually endure another surgery to remove the electrodes. Bleeding within the brain caused by the displacement of electrodes is another dangerous complication that may result.

Infection is a fairly common occurrence (3–10% of cases). Because the risk of intracranial infection is high, the infected electrode must be removed. One solution to the problem is to coat the whole electrode and leads with antibiotics. Skin erosion occurs at the connector site. Low-profile connectors combined with careful placement and closure should reduce the risk. Electrode or wire break may occur when the head is moving or turning, which increases the stress at the connector site. Electrode break requires another intracranial procedure for replacement. Electrode migration is a very serious issue, since it leads to clinical failure. Improvement of fixation techniques may solve this issue. Furthermore, shocking sensations or intermittent stimulation may occur when there is pacemaker dysfunction due, for example, to battery failure. Replacing the battery means another surgical procedure to be performed on the patient, albeit not intracranial.

Beside device malfunctioning, common adverse effects such as psychiatric issues can occur including apathy, hallucinations, personality changes, compulsive behavior, mania, depression, reduction of cognitive functions, psychosis, and even suicide.[19–25]

All the complications with DBS electrodes reported recently in the scientific literature have raised issues that must be addressed before this particular industry segment can achieve complete commercial success. Medical reports indicate that about a quarter of the patients implanted with DBS electrodes have experienced hardware complications such as migration or dislodgement of the leads. The authors tracked specific problems with the electrodes and connectors, which included lead fractures, lead migrations, short/open circuits, erosions, infections and foreign body reactions. Although they reported that up to 70% of the Parkinson's patients experienced notable improvement resulting from the procedure, it is clear that DBS could benefit from further development of the hardware and surgical techniques.[26]

6.3 Motor Cortex Prostheses

Brain-controlled robotic limbs are still far from freeing paraplegics from their wheelchairs or giving amputees their limbs back. These patients do retain the mental dexterity to perform physical actions, even if the spinal cord is severed. By accessing the region of the motor cortex responsible for a specific movement, the intention to move can, in principle, be first decoded and then translated into action with a prosthetic interface. Normally, millions of neurons fire when an arm is lifted. However, the signal generated from only approximately 100 neurons is enough to move a robotic arm at the shoulder, elbow and even clench and open the hand. As a result, the robotic arm becomes an extension of the body that would be easily controlled by thinking.[27]

In the particular case of the peripheral nerves employed for neuroprosthetic applications, the detection and stimulation of their activity using electrodes fabricated by microelectromechanical systems (MEMS) techniques have been an important area of research in neuroengineering for over 25 years. Unlike the electrodes implanted in the brain, which produce significant tissue damage at insertion and later due to the motion of the head, leading to the deterioration of the interface and a limited lifetime, the peripheral electrodes can be used for long times for the detection of the signals and also for the efficient stimulation in order to restore natural sensations with prosthetic limbs. An illustrative example of a biocompatible neural microprobe is presented in Figure 6.6. The microprobes have longitudinal gold electrodes recessed within grooves designed to guide the growth of regenerating axons. A biocompatible polymer (SU8), tested first with primary Schwann cells and explanted root ganglion neurons and then implanted *in vivo*, showed no signs of tissue damage or inflammatory reaction a year after implantation.[28]

Arrays of multisite microelectrodes can be chronically implanted in the brain to collect the intention-driven neuronal activity and convert it into a controlled signal that enables the movement in paralyzed persons who are unable to sense or move their limbs. Figure 6.7 shows an array of 100 electrodes. Each electrode is 1 mm long, spaced 400 µm apart, in a 10×10 grid.[29-33]

A computer–brain interface decodes the signals and transfers the firing patters into motor commands. Subsequently, a computer gateway engages effectors. The applications are intended for people with tetraparesis from spinal cord injury, brainstem stroke, muscular dystrophy or amyotrophic lateral sclerosis. Scalp-based, electroencephalography (EEG) driven brain–computer interfaces, transcranial and electrocorticographic interfaces are currently explored in order to solve the issues related to the implantation surgery and the problems associated with the transcutaneous connections, plus the need to link the patient to bulky equipment that necessitates the constant assistance of a technician.[34-38] Ideally, a wireless and miniaturized system, completely automatic, would be required for practical use, although the challenges to achieve this remain difficult.

For amputated limbs, with intact spinal cord function, electrodes can be implanted only in the residual stump, so the muscle movement modulated by intent is harnessed and then converted into electrical pulses that move the prosthesis. Moving prosthetics at the speed of thought is possible by targeted muscle re-innervation, a technique that implies the cutting of the redundant nerves serving nearby chest muscles that previously moved the missing arm, for example. The motor nerves in the arm stump are then separated and connected to the chest muscles. In four to six months the signals generated from the cortex to move an arm or hand is transmitted to the patient's chest muscles.[39]

The latest advances in robotics technology are extremely promising for partially paralyzed individuals, the disabled and elderly people who have mobility problems due to weak muscles. The 'hybrid assistive limb' is a computerized suit equipped with sensors that detect the brain signals produced when a person attempts to move. The sensors relay these signals to the motors and computer inside the suit, which work together to direct limb movement through mechanical leg braces strapped to the thighs and knees. All this happens with only fractions of a second delay. The system provides the individual with motor assistance. The 10 kg computer is wireless, battery operated and belted to the waist. The suit can not only help the disabled and elderly, but could also be used to aid caregivers lift or move those who are ill or infirm, and even help laborers to move heavy equipment. The system is an example of a

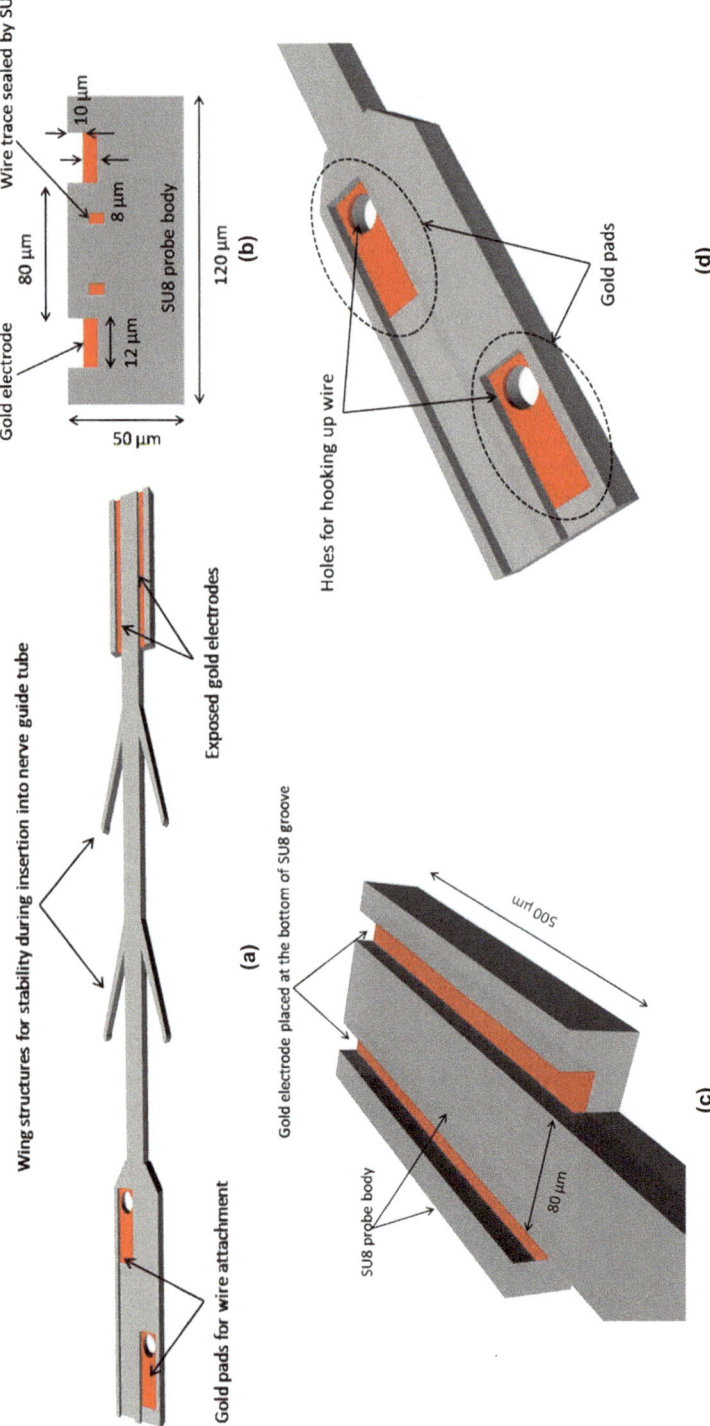

Figure 6.6 Design of the neural microprobe. (a) Top view of major microprobe features including gold wire bonding pads (left), flexible centring 'wings' along microprobe shaft and gold 'groove' electrodes (right). (b) Cross-section of grooved electrode channel structure. (c) Top view of grooved electrodes. (d) Gold wire bonding pads. Bipolar gold electrodes along the floor of the grooves are 500 μm long and 12 μm wide.[28] (Reprinted with kind permission of the IEEE.)

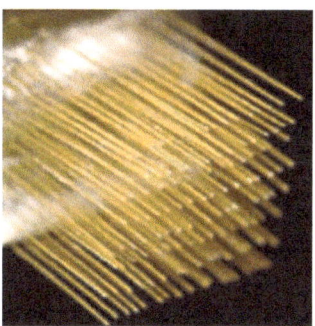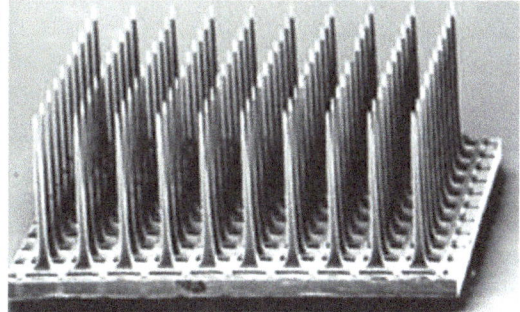

Figure 6.7 Microfabricated electrode arrays for brain prostheses.[29] (Reprinted with the kind permission of Nature Publishing Group.)

new field known as 'cybernics' that incorporates technological and mechanical elements from the areas of mechanics, bionics, electronics and robotics.[40,41]

In order to detect signals originating in the brain, the structure depicted in Figure 6.8 suggests a way to overcome difficulties related to the signal intensity evanescence in time by using electrodes that sense the location of the strongest neuronal signal and consequently move towards that location. The whole device is mounted on the skull and employs piezoelectric micromotors to four electrodes, independently of each other, in one micron increments. In order to prevent damage to the neurons, the device has a collision avoidance capability that restricts the movement when the voltage rises rapidly indicating there is danger of puncturing a neuron. Such implants can decode motor signals in rats and intention signals in monkeys, and can be fitted onto paralyzed individuals allowing them to control a computer cursor and navigate the web.[42,43]

6.4 Replacing Damaged Brain Components

Scientists not only hope to stimulate brain activity with electronic devices, but also to replace damaged parts of the brain in individuals who suffer from epilepsy or Alzheimer's disease and those with brain damage due to strokes. There are deep ethical concerns that surround the use of devices that mimic brain activity. These concerns are motivated by the potential effects of such a device on memory, mood, awareness and consciousness, all of which contribute to an individual's fundamental identity. Although further research is required to investigate whether or not brain implants could influence an individual's personality, high level cognitive brain signals can be decoded with implants. Of course the precise mechanism through which the brain encodes information remains elusive. Accordingly, the only strategy open at this point is a low level of functional mimicry. For example, slices of animal brain can be stimulated with electrical signals millions of times over until a pattern of inputs and outputs can be established. Then the information from various slices can be functionally assembled together, resulting in a mathematical model of a specific part of the brain. It sounds deeply mechanistic and reductionistic in character,

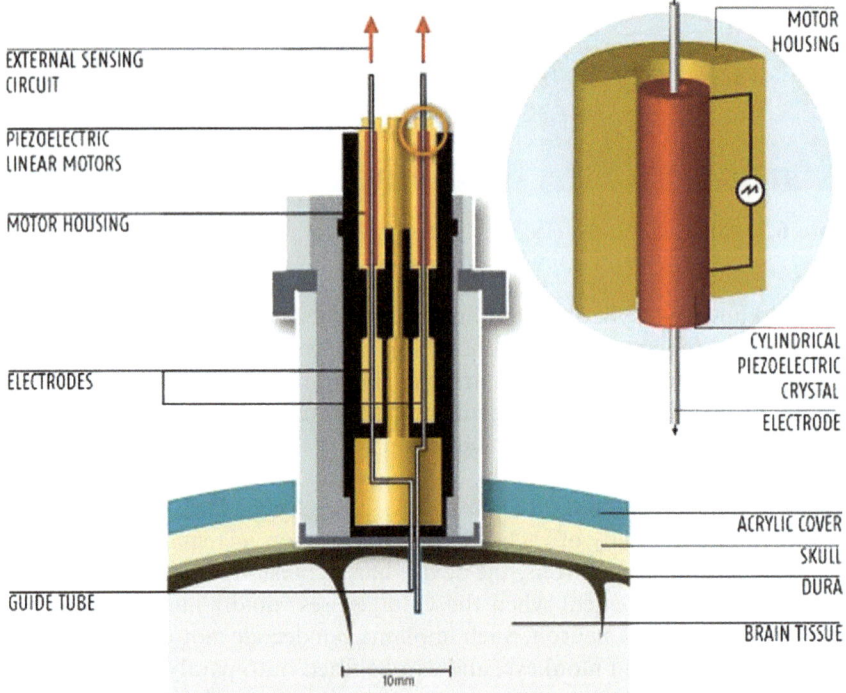

Figure 6.8 Structure of a mobile scalp electrode system able to detect and move towards the location of the strongest neuronal signal while avoiding the puncture of neurons.[42]
(Reprinted with the kind permission of the IEEE.)

but if specific networks of neurons are isolated, the approach might work. Because the hippocampus is the most ordered and structured part of the brain and one of the most studied, it would be meaningful to test its function first and develop a mathematical model. Later, such models and biomimetic devices could be translated to other regions of the human brain.

A chip containing the program can be envisioned to sit on the skull rather than inside the brain. Communication with the brain would be achieved through two electrode arrays placed on the either side of the damaged area. As shown in Figure 6.9, the first array records the input electrical activity originating from the neuron and re-routes the signal to a programmed chip. This bypasses the damaged region of the brain. The chip then processes the information and transmits the signal to the second electrode array, where the signal can be communicated to the rest of the brain.

The Biomimetic Interface between Brain and Electrodes

Of course, the brain does not function like computer hardware. However, such measurements might shed light onto the way information is processed between different levels and areas of the brain. Even after all the recent brain research assisted by computed axial tomography (CAT) scans, positron emission tomography (PET) scans and functional magnetic resonance imaging

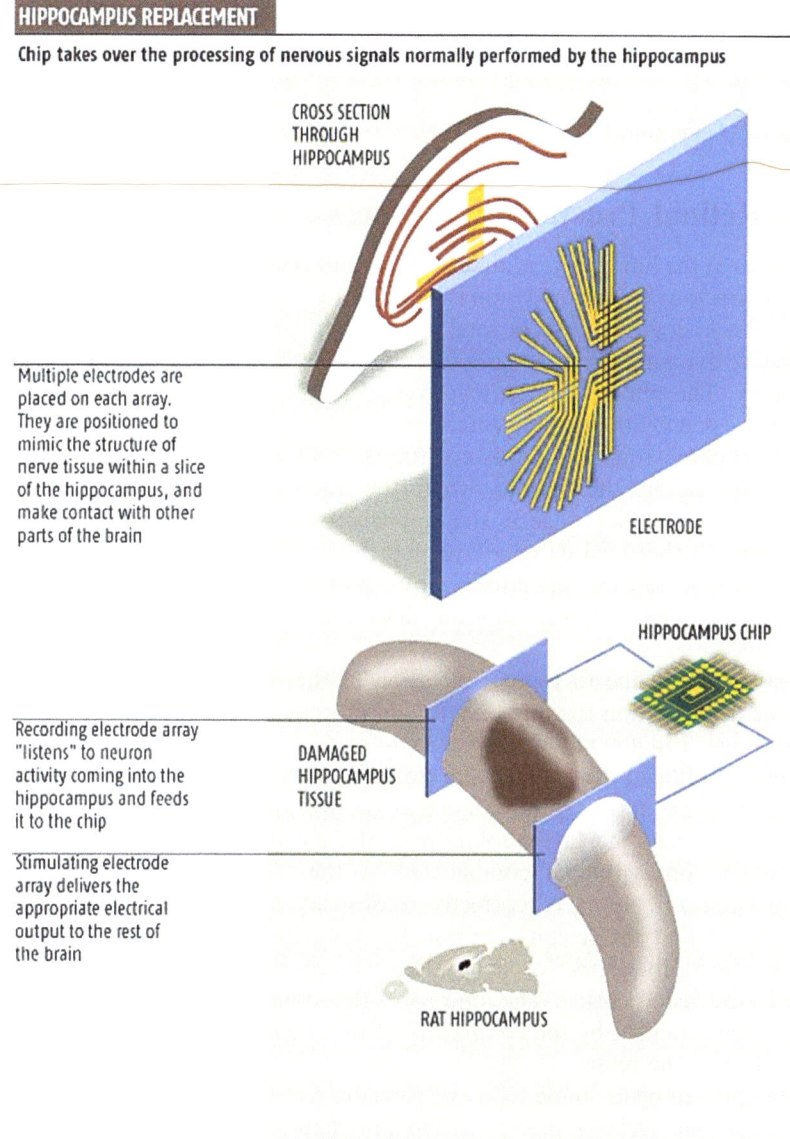

Figure 6.9 Shortcutting damaged neuronal tissue by double arrays of electrodes.[43] (Reprinted with the kind permission of the IEEE.)

(fMRI), neuroscientists still don't know how diverse perceptions occurring in distant parts of the brain are bound together in a single conscious experience. The state-of-the-art research plan in neurobiology is still to find the neurons or subcircuits of interconnected neurons, their feedback and inhibitory mechanisms that would explain brain activities such as thinking, emotions, intuition, creativity and insight. Neuroscientists are looking for neural correlates, presuming that the rate and manner of firing correlate with conscious experiences. The search is focused on finding the neural code and to discover what exactly is associated with experiences, the mean firing rates or the chemical behavior of neurotransmitters. Nevertheless, the complexity of the brain might exist in a much higher dimension, at an unthinkable and inconceivable frontier.

6.5 Retinal Prosthetic Interfaces

The retina in the human eye is an impressive and very fragile tissue on which the whole marvelous process of vision is based. It receives and converts the incident light, be it a single photon in total darkness to the brightest sunlight, on an impressive dynamic range, remarkable color discrimination and with exquisite sensitivity. The photostimuli are converted in the retina to a signal which is later processed by the visual centers of the brain. Millions of photoreceptors (100–150 million rods and cones creating a condensed network in the central area of the retina, the macula, where the resolution is maximal and vision sharpest) convert the image in the language of neurons in the bipolar layer, connecting to the third ganglia layer which transfers the signals to the nerve fiber layer that transfers the information—now reduced by a factor of 100, and processed to account for motion detection and edge enhancement—to the brain.

Loss of vision due to the degeneration of the retina occurs in age-related macular degeneration (people over 65), in retinitis pigmentosa (occurring at a younger age, 1 in 4000 live births), and numerous other medical conditions or accidents. Artificial means of restoring lost vision cannot replace millions of optical receptors in the retina, situated at about 5 μm distance from each other. Nor can it attain the fine resolution of the macula. Moreover, the task of restoring vision is further complicated by the fact that the brain requires multiple sensory inputs to correctly reconstruct a three-dimensional image. This implies that the sensory neurons be stimulated not only in parallel, but also in a spatially correct order to allow an accurate image encoding to take place in the visual cortex. The impressive functional intricacies of the eye are further compounded by the complexity of the brain's interpretation of signals originating in the retina.

Attempts are being made today to partially restore lost vision using current microelectronic technologies as visual prosthetics. These prostheses can be connected directly to the brain or implanted in the eye if the optic nerve is still intact. Presently, there are three strategies in the technical development for visual prosthetics—cortical implants, optic nerve stimulation and retinal implants. In the case of electrodes implanted directly in the visual cortex, pain around the implant occurs due to the irritation of the meninges by the electrode current. In order to alleviate this pain, smaller electrodes and reduced currents must be used. The main disadvantage of cortical implants is the risk of surgical complications during the implantation procedure. The advantages of this strategy are that the electrodes are protected by the skull and the possibility of completely bypassing the damaged retina or optic nerve. Placing electrodes around the optic nerve allows the stimulation of the visual cortex using a video device situated outside the brain. This strategy requires the optic nerve to be intact. However, there is a risk of infection from the external wires.

With regard to retinal prostheses, there are mainly three types of approaches that can be employed. These are subretinal microphotodiode arrays, epiretinal and subretinal electrode arrays.[44] As a result of the direct contact of the electrodes with such a sensitive neural tissue, biocompatibility considerations become extremely important. Complete hermeticity of the encapsulated intraocular components and special materials that do not corrode or release toxic contaminants must be used.[45] The implanted electrodes must be covered by a capacitive layer with very high resistance. Tissue reactions from mild formation of loose vascularized tissue to newly formed tissue covering the electrodes raise the impedance, which strains the already challenged power supply. Since the retina itself is approximately 250 μm thick, such growth cannot be tolerated. The central vision is given by the macula area of only 1–2 mm in diameter, usually the first target of the degenerative processes. The presence of the prostheses itself can speed up the deterioration by limiting the nutrient and the oxygen supply to the retina. In particular, the tissue reaction involves a cascade of events at the molecular level, mainly the production and release of neuroinflamatory cytokines—interleukins, tumor necrosis factor-α (TNF-α), monocyte chemotactic protein-1 (MCP-1), platelet-derived growth factor (PDGF).[46] Electrochemical impedance spectroscopy methods are able to detect and quantify the process with high accuracy.[47,48] Finally, retinal detachment, cataract formation, changes in the retinal pigment endothelium, infection and local inflammation due to surgical insertion are all potential risks which can endanger the peripheral vision that patients still retain.[49,50]

Scientists around the world are trying to circumvent all these apparently insurmountable difficulties and restore at least partial functional vision to the blind in order for them to regain some independence and at least spatial orientation and feature recognition for everyday life. Epiretinal arrays of microelectrodes can be connected to an exterior camera that captures the images and subsequently converts them into an electrical stimulation pattern imposed on the remaining functional retina. The array is essentially a readout chip coupled to the ganglion cells. The extraocular unit is sutured to the outside of the eye onto the sclera. Radiofrequency transmitter coils for power and data are mounted on the patient's glasses and receiving coils are implanted on the temporal side of the eye.[51]

These arrays have the main advantage of dissipating the heat from the electronic components into the vitreous. But because edema can develop under the array and thicken the retina, retinal microtacks must be used to secure the array in place (glial cells replacing retinal cells under the tacks). Because these tacks can break as a result of eye movement, subretinal implantation is a more desirable strategy. Subretinal arrays are implanted into the space that would be occupied by photoreceptors in a healthy retina between the pigment epithelial layer and the outer layer of the retina. Unlike epiretinal implants, the remaining neural network of the retina is used, without any need of extra signal processing. The implant is held in place by natural ocular pressure and thus avoids the need for microtacks. There is direct contact with the neural retina. This allows the natural neural circuitry to carry out the signal processing and therefore makes it the most efficient neural–microelectrode interface. The close proximity to the bipolar cells allows for less intense stimulation and thus minimizes the destruction of the sensitive retinal layer.

A variation of the subretinal implant is the microphotodiode array. These are known as artificial silicon retina because they mimic the function of natural photoreceptors. The silicon array utilizes the same principles that govern the conversion of solar energy to electrical energy in solar cells.[52] For instance, when light reaches the array of photodiodes, photons within the diodes are converted into electrical signals that locally stimulate neuron cells of the retina. There is no need for an external power source since the incident light is, in principle, sufficient to power the array. However, since existing photodiodes are relatively inefficient, approximately a thousand times less efficient than natural receptors, only intense light can provide sufficient photon flux. Moreover, the *in vivo* stability of the silicon is poor and this results in the progressive degradation of the microarray. Severe corrosion completely dissolves the silicon oxide passivation layer within 6–12 months after implantation. Once the passivation layer is degraded, the bulk silicon underneath this layer begins to corrode. The exact mechanism that is responsible for this extensive corrosion is unknown. It may be due to local electrochemical processes and the reaction of strong oxidase enzymes in macrophage cells present at the implant site. The difficulties associated with achieving satisfactory biocompatibility and thus long-term implant stability must be overcome before these prostheses can be made available to the visually impaired.

Alternative microphotodiode arrays constructed from ceramic materials and based on the photoferroelectric effect are being studied[53] in order to circumvent the problem of silicon degradation, although a fully functional artificial retinal prosthetic has not yet been achieved. Systems currently being researched require general anesthesia and a six-hour operation to implant surgically, construct and connect multiple pieces of hardware in the eye, or alternatively, to insert surgically an implant into the eye which is connected to a wire passing through the patient's skull. Patients wear eyeglasses with an external camera and transmitter as well as a belt with a video processor and battery that charges the system. The patient is able to see forward, but must move the head to change the field of view. These systems provide up to 60 pixels of sight capacity,

Figure 6.10 Penetrating electrodes designed to replace the damaged photoreceptor in the eye.
(Image kindly supplied by Nano-Retina Inc., Israel.)

i.e. a patient can differentiate between dark and light and perhaps identify the existence of an object. Figure 6.10 shows a 5000 pixel retinal implant. A rechargeable, battery-powered mini laser, situated on a pair of eyeglasses, efficiently powers the implant wirelessly. However, research efforts[54,55] have shown more promise than alternative treatments such as gene therapy to replace defective genes in photoreceptors, laser photoirradiation to destroy the neovascularization that causes wet macular degeneration, transplantation, and the use of anti-apoptotic agents to delay the degeneration and death of photoreceptor cells.

6.6 Technological Advances and Novel Strategies for Improving the Electrode–Brain Interface

Replacing bulky electrodes with nanowires threaded through blood vessels might one day be the solution to avoid side effects related to the insertion of electrodes. Platinum nanowires guided through blood vessels can be used to detect the activity of individual neurons.[56] The difficulties related to guiding the wires through fragile branches of the brain's vascular system can be solved using new conductive polymers nanowires, which change the shape in response to variable electrical fields, which may allow the steering of probes. Polymer nanowires are 20–30 times smaller than the platinum ones and they can be made biodegradable for short-term brain implantation.

Soft, fuzzy biomaterials and electroactive conductive polymers—poly(3,4-ethylenedioxythiophene) poly(styrenesulfonate) (PEDOT), biologically conjugated melanin, *etc.*—can also be deposited on any kind of electrodes to improve their biocompatibility.[57] However, polymer materials are problematic, especially in long-term biocompatibility, as they are not sealed completely and become conductive. Sputtered iridium oxide film significantly improves the neural interface devices. The film exhibits better electrochemical characteristics in comparison to platinum. Its charge injection capacity is an order of magnitude higher than that of Pt, while impedance (at 1 kHz) is ten times lower than that of platinum, making it an attractive stimulation and recording electrode material.[48] Carbon nanotubes are a promising solution for implantable electrodes. They express metallic or semiconductive properties depending on the folding modes of the nanotube walls and represent a novel class of nanowires. Different methods to separate semiconductive from conductive carbon nanotubes have been developed and synthetic strategies to chemically modify the side walls or tube ends by molecular or biomolecular components have been reported. Tailoring such hybrid systems consisting of carbon nanotubes and biomolecules (proteins and DNA) has expanded rapidly and attracted substantial research effort. The same integration of biomaterials enables the use of the hybrid systems as active field-effect transistors (FETs) or biosensor devices, and allows the generation of complex nanostructures and nanocircuitry of controlled properties and functions, with potential applications in nanobioelectronics and nanobiotechnology.[58,59]

Optical neural interface might be an alternative possibility for solving the debilitating side effects of DBS or the lack of specificity in targeting excitatory or inhibitory neurons.[60] Integrated fiber optic and laser diode optobionic systems may also constitute a more natural interface for retinal prostheses, based on the photostimulation of retinal neurons. Typically $100\,\text{mW}\,\text{cm}^{-2}$ in instantaneous light intensity is necessary to stimulate action potentials in neurons. This can be reduced to safe levels in order to negate thermal and photochromic damage to the eye.[61]

Neural implants have progressed from wires to micro- and nanowires, to silicon arrays and to fiber optic interfaces. The next generation will be represented by flexible structures with engineered surfaces and bioactive coatings for a better integration with the neural tissue. To insert them, dissolvable coatings can be applied to stiffen the electrode. The coating can rapidly disintegrate after insertion. Such electrodes can be equipped with amplifiers and circuits for magnetic or thermal stimulation. Micromachined channels could deliver drugs at specific sites in the brain. For long-term applications such implants must use advanced biomaterials to prevent the foreign body effect expressed by local inflammation, protein absorption, cellular adhesion and formation of the fibrous scar tissue. Special coatings and immobilization of active biological molecules can be designed to either mimic an extracellular matrix or provide cell binding domains. Stimulation of natural neuroplasticity in the brain using such active implants represents an exciting possibility. Only intense interdisciplinary research crossing engineering, chemistry, material science and neurobiology will be able to answer such demands.

References

1. J. P. Boehmer, *Am. J. Cardiol.*, 2003, **91**, 53D.
2. L. Cheran, M. Gheorghe and D. Dascalu, *Innov. Tech. Biol. Med.*, 1995, **16**, 758.
3. L. I. Terr, J. P. Mobley and W. F. House, *Am. J. Otol.*, 1989, **10**, 339342.
4. L. I. Terr, G. A. Sfogliano and S. L. Riley, *Laryngoscope*, 1989, **99**, 11711174.
5. D. H. Kim, S. Richardson-Burns, L. Povlich, M. R. Abidian, S. Spanninga, J. L. Hendricks and D. C. Martin, in *Indwelling Neural Implants: Strategies for Contending with the In vivo Environment*, ed. W. M. Reichert, CRC Press, Boca Raton, FL, 2008, 177–220.
6. V. Polikov, M. Block, C. Zhang, W. M. Reichert and J. S. Hong, in *Indwelling Neural Implants: Strategies for Contending with the In vivo Environment*, W. M. Reichert, CRC Press, Boca Raton, FL, 2008, 89–116.
7. P. D. Wolf, in *Indwelling Neural Implants: Strategies for Contending with the In vivo Environment*, ed. W. M. Reichert, CRC Press, Boca Raton, FL, 2008, 63–88.
8. W. M. Grill, in *Indwelling Neural Implants: Strategies for Contending with the In vivo Environment*, ed. W. M. Reichert, CRC Press, Boca Raton, FL, 2008, 41–62.
9. W. He and R. V. Bellamkonda, in *Indwelling Neural Implants: Strategies for Contending with the In vivo Environment*, ed. W. M. Reichert, CRC Press, Boca Raton, FL, 2008, 151–176.
10. M. J. Bridge and P. A. Tresco, in *Indwelling Neural Implants: Strategies for Contending with the In vivo Environment*, ed. W. M. Reichert, CRC Press, Boca Raton, FL, 2008, 117–150.
11. J. J. Fins, N. D. Schiff and K. M. Foley, *Neurology*, 2007, **68**, 304307.
12. N. D. Schiff and J. J. Fins, *Curr. Opin. Neurol.*, 2007, **20**, 638642.
13. N. D. Schiff, J. T. Giacino, K. Kalmar, J. D. Victor, K. Baker, M. Gerber, B. Fritz, B. Eisenberg, T. Biondi, J. O'Connor, E. J. Kobylarz, S. Farris, A. Machado, C. McCagg, F. Plum, J. J. Fins and A. R. Rezai, *Nature*, 2007, **448**, 600603.
14. P. A. Tass, *Biol. Cybern.*, 2003, **89**, 81.
15. C. R. Butson and C. C. McIntyre, *Clin. Neurophysiol.*, 2005, **116**, 2490.
16. C. R. Butson, C. B. Maks and C. C. McIntyre, *Clin. Neurophysiol.*, 2006, **117**, 447.
17. X. F. Wei and W. M. Grill, *J. Neural Eng.*, 2005, **2**, 139.

18. J. L. Shils, L. Z. Mei and J. E. Arle, *Stereotact. Funct. Neurosurg.*, 2008, **86**, 16.
19. C. C. Luca and C. Singer, *Parkinsonism Relat. Disord.*, 2013, **19**, 466.
20. C. Fukaya, K. Shimoda, M. Watanabe, T. Morishita, K. Sumi, T. Otaka, T. Obuchi, K. Toshikazu, K. Kobayashi, H. Oshima, T. Yamamoto and Y. Katayama, *Neuromodulation*, 2012, DOI: 10.1111/j.1525-1403.2012.00516.x..
21. J. L. Ostrem, N. B. Galifianakis, L. C. Markun, J. K. Grace, A. J. Martin, P. A. Starr and P. S. Larson, *Clin. Neurol. Neurosurg.*, 2012.
22. A. Eusebio, H. Cagnan and P. Brown, *Front. Integr. Neurosci.*, 2012, **6**, 47.
23. A. Franzini, R. Cordella, G. Messina, C. E. Marras, L. M. Romito, A. Albanese, M. Rizzi, N. Nardocci, G. Zorzi, E. Zekaj, F. Villani, M. Leone, O. Gambini and G. Broggi, *Neurol. Sci.*, 2012, **33**, 1285.
24. K. Witt, C. Daniels and J. Volkmann, *Parkinsonism Relat. Disord.*, 2012, **18**(Suppl 1), S168.
25. U. Sandvik, G. M. Hariz and P. Blomstedt, *Acta Neurochir.*, 2012, **154**, 495–499.
26. M. Y. Oh, A. Abosch, S. H. Kim, A. E. Lang and A. M. Lozano, *Neurosurgery*, 2002, **50**, 1268; discussion 1274.
27. A. Bragin, J. Hetke, C. L. Wilson, D. J. Anderson, J. Engel, Jr. and G. Buzsaki, *J. Neurosci. Methods*, 2000, **98**, 77.
28. S. H. Cho, H. M. Lu, L. Cauller, M. I. Romero-Ortega, J. B. Lee and G. A. Hughes, *IEEE Sens. J*, 2008, **8**, 1830.
29. L. R. Hochberg, M. D. Serruya, G. M. Friehs, J. A. Mukand, M. Saleh, A. H. Caplan, A. Branner, D. Chen, R. D. Penn and J. P. Donoghue, *Nature*, 2006, **442**, 164.
30. L. R. Hochberg and J. P. Donoghue, *IEEE Eng. Med. Biol.*, 2006, **25**, 32.
31. L. R. Hochberg, M. D. Serruya, J. A. Mukand, G. M. Friehs and J. P. Donoghue, *Neurology*, 2006, **66**, A101.
32. G. Friehs, R. D. Penn, M. C. Park, M. Goldman, V. A. Zerris, L. R. Hochberg, D. Chen, J. Mukand and J. D. Donoghue, *Neurosurgery*, 2006, **59**, 481.
33. A. Kubler, V. K. Mushahwar, L. R. Hochberg and J. P. Donoghue, *IEEE Trans. Neural Syst. Rehab. Eng.*, 2006, **14**, 131.
34. P. Kennedy, D. Andreasen, P. Ehirim, B. King, T. Kirby, H. Mao and M. Moore, *J. Neural Eng.*, 2004, **1**, 72.
35. E. C. Leuthardt, G. Schalk, J. R. Wolpaw, J. G. Ojemann and D. W. Moran, *J. Neural Eng.*, 2004, **1**, 63.
36. B. Graimann, J. E. Huggins, S. P. Levine and G. Pfurtscheller, *IEEE Trans. Biomed. Eng.*, 2004, **51**, 954.
37. N. Jiang, S. Muceli, B. Graimann and D. Farina, *Med. Biol. Eng. Comput.*, 2012, 143–151.
38. J. M. Hahne, B. Graimann and K. R. Muller, *IEEE Trans. Biomed. Eng.*, 2012, **59**, 1436.
39. L. A. Miller, K. A. Stubblefield, R. D. Lipschutz, B. A. Lock and T. A. Kuiken, *IEEE Trans. Neural Syst. Rehabil. Eng.*, 2008, **16**, 46.

40. K. Suzuki, M. G., H. Kawamoto, Y. Hasegawa and Y. Sankai, *Adv. Robot.*, 2007, **21**, 1441.
41. S. Musallam, B. D. Corneil, B. Greger, H. Scherberger and R. A. Andersen, *Science*, 2004, **305**, 258.
42. R. Huang, C. Pang, Y.-C. Tai, J. Emken, C. Ustun, D. C. Rizzuto, R. A. Andersen and J. W. Burdick, in *Proceedings 21st IEEE International Conference on Micro Electro Mechanical Systems (IEEE-MEMS'08)*, IEEE, New York, 2008, pp. 240–243.
43. C. Pang, J. Cham, Z. Nenadic, S. Musallam, Y.-C. Tai, J. Burdick and R. Andersen, *Conf. Proc. IEEE Eng. Med. Biol. Soc.*, 2005, **7**, 7114.
44. C. Sekirnjak, P. Hottowy, A. Sher, W. Dabrowski, A. M. Litke and E. J. Chichilnisky, *J. Neurosci.*, 2008, **28**, 4446.
45. C. Scholz, *J. Bioact. Compat. Polym.*, 2007, **22**, 539.
46. D. L. Vargas, C. Nascimbene, C. Krishnan, A. W. Zimmerman and C. A. Pardo, *Ann. Neurol.*, 2005, **57**, 67.
47. W. Liao and X. T. Cui, *Biosens. Bioelectron.*, 2007, **23**, 218.
48. S. Negi, R. Bhandari, L. Rieth and F. Solzbacher, *Biomed. Mater.*, 2010, **5**, 015997.
49. E. Zrenner, *Science*, 2002, **295**, 1022.
50. D. Besch and E. Zrenner, *Doc. Ophthalmol.*, 2003, **106**, 31.
51. C. de Balthasar, S. Patel, A. Roy, R. Freda, S. Greenwald, A. Horsager, M. Mahadevappa, D. Yanai, M. J. McMahon, M. S. Humayun, R. J. Greenberg, J. D. Weiland and I. Fine, *Invest. Ophthalmol. Vis. Sci.*, 2008, **49**, 2303.
52. A. Y. Chow, V. Y. Chow, K. H. Packo, J. S. Pollack, G. A. Peyman and R. Schuchard, *Arch. Ophthalmol.*, 2004, **122**, 460.
53. E. D. Flinn, *Aerospace Am.*, 2002, **40**, 18.
54. A. Horvat-Broecker, J. Reinhard, S. Illes, T. Paech, G. Zoidl, S. Harroch, C. Distler, P. Knyazev, A. Ullrich and A. Faissner, *Neuroscience*, 2008, **152**, 618.
55. T. C. Pappas, W. M. S. Wickramanyake, E. Jan, M. Motamedi, M. Brodwick and N. A. Kotov, *Nano Lett.*, 2007, **7**, 513.
56. R. R. Llinas, K. D. Walton, M. Nakao, I. Hunter and P. A. Anquetil, *J. Nanopart. Res.*, 2005, **7**, 111.
57. M. S. Shoichet, C. C. Tate, M. D. Baumann and M. C. LaPlaca, in *Indwelling Neural Implants: Strategies for Contending with the In vivo Environment*, ed. W. M. Reichert, CRC Press, Boca Raton, FL, 2008, 221–244.
58. N. Sinha and J. T. Yeow, *IEEE Trans. Nanobiosci.*, 2005, **4**, 180.
59. E. Katz and I. Willner, *Chemphyschem*, 2004, **5**, 1085.
60. A. R. Adamantidis, F. Zhang, A. M. Aravanis, K. Deisseroth and L. de Lecea, *Nature*, 2007, **450**, 420.
61. P. Degenaar, N. Grossman, M. A. Memon, J. Burrone, M. Dawson, E. Drakakis, M. Neil and K. Nikolic, *J. Neural Eng.*, 2009, **6**, 035007.

CHAPTER 7

A Look at the Future

The detection configurations presented in this book constitute a clear demonstration of the intimate integration of physics, electrical and biomedical engineering, bioanalytical chemistry, surface physical chemistry and advanced technologies in biology and medicine. This is highly reflective of the emergence of the relatively new paradigm of integrative science where boundaries between traditional areas of science, medicine and engineering disappear. Research based on such advanced detection systems offers enormous potential for bridging the chasm between technology and biology through the provision of methods and devices that are capable of the elucidation of new insights into biological processes. Such new knowledge will undoubtedly lead to enhanced approaches to the practice of medicine, especially as it pertains in this case to areas of the discipline such as neurology.

7.1 Quantum Neurobiology

It has been said by scientists that the 20th century was the era of progress in physics as extraordinary advancements were made, especially in the key realm of quantum physics. The 21st century is predicted to be the era of biology, where technological developments are expected to enable the connection between molecular biology and this area of physics. The merging of quantum physics and biology, known as quantum biology, shows great promise in potentially shedding light on many unanswered questions in biology, medicine, psychology and neuroscience, and deeply challenging topics such as the understanding of consciousness. There is no doubt that expansion of quantum biology in the coming research years will lead to a number of surprises, not only in biology, but also in fields such as neurochemistry. As implied in the quotes at the very beginning of this text, completely new approaches may be required in order, for example, to understand the true nature of human consciousness.

Certainly the relatively new science of quantum biology may represent such a new strategy in elucidation of brain function where many questions remain unanswered.

The brain is a near, but still such a distant frontier to be explored. Science is unable to explain how mental processes such as creativity, intuition, insight, thinking, emotions and the feeling of being alive arise from electrochemical impulses firing along neural axons. Despite the elegance of quantum physics and its mathematical nature, it is still difficult to view its connection to conscious behavior. Neuroscientists can see the activity in distinct areas of the brain, but are at a loss to explain how these are bound together in unitary experiences and feelings. How can a simple thought initiate such a cascade of neural activity inside the brain? How can the nonmaterial mind influence the material brain? Is consciousness the substrate of both, conceiving, forming and becoming biology, so interaction happens naturally? Even if, for now, science has more questions than answers, we should never stop questioning but instead always use the power of our curiosity and imagination.

7.2 Nanoneuromedicine

Advances in micro- and nano-scale technology, high-speed signal processing and the enormous memory storage of electronic devices will enable the development of hybrid bio-electronic systems capable of performing fundamental studies at the molecular level. Progress in neuroscience depends on highly sensitive investigational devices and the future of regenerative medicine may well be decided by the potential of nanosystems and related technologies. We already witness the integration of neurobiology with microtechnology due to the ideal match in terms of dimensionality with the biological components represented by living neuron cells. This has resulted in many exciting applications in neurobiology, diagnostics, drug discovery and prosthetic implant technology, and new developments are expected in neuron-based processors for biocomputers or implantable information processing devices for virtual reality interfacing.

Understanding of disease at the molecular level is the cornerstone of nanomedicine. It goes beyond a mechanistic view of the human body, probing the fundamental biochemical, electrical and energetic properties of living tissue through the utilization of technology at the molecular scale. The current trend of monitoring disease biomarkers using modern bioanalytical methods illustrates the progress in this field. Tracking biomarkers enables swift and accurate diagnoses so that the appropriate treatment can be implemented before the disease progresses and causes further physiological damage. Future nanomedicine will be concerned with improving the efficacy of current pharmaceutical therapies. 'Smart' microfluidic systems can be used to optimize drug delivery and minimize patient discomfort, while nano-agents are developed to allow medical treatments to traverse the blood–brain barrier.

7.3 Neuropharmacology

Nanotechnologies target the brain at the molecular scale. A particular application of such new technologies is the investigation of the effect drugs have on neurons with particular regard to regenerative purposes. Regenerative medicine shows remarkable promise not only to treat symptoms but also to restore neural functionality, both at the peripheral and at the central nervous system level. An important aspect of this is the investigation of neuron regeneration and neuron plasticity, with a special focus on neurogenesis, the production of nerve cells from stem cells. The generation of neural stem cells happens deep within the brain, where they subsequently migrate to become part of the circuitry of the brain. Neuro-electronic devices have tremendous potential in evaluating the best course of action for the use of stem cells in the re-population or replacement of brain cells destroyed by injury or disease.

A number of experimental drugs are thought to be effective in the treatment of neurodegenerative conditions but, for the most part, their mechanism of action remains largely unknown. Furthermore, finding the correct drug and its proper dosage often relies on trial and error. Therefore, these diseases can continue a slow progression and are difficult to detect. In addition, specific drug responsiveness depends strongly on the individual patient, thus requiring considerable personalization, even if the molecule in question is considered capable of crossing the brain–blood barrier. Treating the symptoms and slowing down further degeneration is often all that can be achieved today for such diseases.

7.4 Genetics

As we entered the 21st century many thought that the greatest hope for the practical outcome of the Human Genome Project was the eventual elimination of genetic diseases, including those of the neural variety. The overriding concept was that certain diseases could be stopped 'at the source'. After spending many millions of dollars to decode the human genome, the Holy Grail was not found: the contribution of the genes variants to disease is rather weak. When it comes to genetics and the brain there have been attempts to relate genes to the likes of criminality and anti-social behaviour, with the expected limited success.

Human genetic engineering involving the addition or deletion of defect genes in the genetic material provided by parents is largely resisted on ethical grounds. Somatic gene therapy performed by manipulating the genome of an individual in order to obviate deleterious effects of a single gene gave unfounded hopes and generated extreme technical difficulties even in the case of very well researched genetic conditions such as the diseases of the brain or central nervous system, Huntington's disease and multiple sclerosis. Beyond that, the problems are compounded by the fact that all common diseases involve not just a single gene but a concert of genetic mutations. Recent microarray technologies can now identify thousands of gene variants and very soon the ability to read the entire 25 000 genes of a person will become a great tool in prospective diagnostic, while generating agonizing moral dilemmas for parents who have to decide on how to act knowing that the future child will be affected by a genetic disease.

Identifying the genes associated with neurological disorders may lead to a better understanding of the contribution of a specific gene, the role of proteins and related biochemical mechanisms, with practical application in pharmacogenetics and the development of psychopharmaceuticals. A good example of using high sensitivity methods to detect the effects of drugs on neurons is illustrated in this book in Chapter 5, where the behaviour of cells in resumes to neurotrophic factors is discussed. In this work involving acoustic wave detection it was shown that planar populations of neurons are indeed affected by such experimental drugs, but the link between the physic and cellular characteristics remains obscure. Even more challenging is the question of how relevant this type of research is to medicine, and neurology in particular. The jury is out with respect to this issue. Currently, neurological disorders are difficult to treat and diagnosis, as specific biomarkers that identify distinct patient populations and their response prognosis are not yet available. It has been seen with current psychiatric disorders that, while one patient may have a complete, positive response, another person with the very same diagnosis may have a terrible response with horrible side effects. This suggests the need for genetic correlation and the development of new methods to treat such diseases in a more personalized way, with targeted psychopharmaceutical treatments.

As a final remark in this section it is unsurprising to note that proteome and proteomics, where the concentration is on the products of gene expression rather than genes themselves, is still an extremely active and growing area of research.

7.5 Cognitive Enhancers

Research aimed at arresting or improving the cognitive decline that occurs in conditions such as Alzheimer's disease, or the emotional mood in depression, is extended in current times to healthy individuals. This includes a wide cross-section of the community such as students cramming for exams, older individuals concerned with loss of memory, and others concerned with their overall level of intellectual 'performance'. Drugs such as Aricept and rivastigmine, designed to boost acetylcholine transmission in Alzheimer's patients or memantine, which affects glutamate neurotransmission, have been seen as possible cognitive enhancers. There are adverse side effects connected with the use of these drugs and, accordingly, it is doubtful if they will be employed generally as cognitive enhancers.

The elucidation of the molecular mechanisms involved in synaptic modulation during memory formation might reveal molecular alternatives. Some possibilities would be substances that mimic the normal role of the amyloid precursor protein, or neuromodulators such as brain-derived nerve growth factor. Another option would be represented by substances that turn on the genes regulating the protein synthesis for memory formation or molecules that enhance the efficacy of the glutamate receptors. Blocking the glutamate receptors in an attempt to selectively erase memory and alleviate posttraumatic stress disorder symptoms is another direction for neuroscience research. One aspect of this sort of approach that must be considered in the future is the ethical ramifications. This is always going to be the case when it comes to drugs used for the treatment of social disorders such as aggression or oppositional defiance.

7.6 Brain Imaging

Functional magnetic resonance imaging (MRI) is an extraordinary tool employed for the detection of sites of brain hemorrhage, stroke, tumors and injury. Even if there are not always efficient treatments available to heal brain diseases or indeed injuries, a fast and precise diagnostic tool is essential to aid in the rendering of immediate medical decisions and to help determine prognostic factors. Recent years have seen MRI technology being applied to the diagnosis of functional deficits in reasoning, memorization, personality disorders and psychopathology. The method involved is the monitoring of brain areas that 'light up' differentially in affected patients. Extended application to the more controversial areas such as the employment of imaging detection in sorting normal people from dangerous criminals and terrorists, while well justified, might open a Pandora box of social abuse. Yet another controversial application is the use of MRI as an alternative to the classical but unreliable lie detector test, which is based on skin impedance variations.

7.7 Stem and Cancer Cells

It is hoped that introduction of the techniques discussed in this text, and other approaches such Raman and MRI imaging, will also prove very fruitful in stem cell and cancer cell research. The current use of stem cells and cell therapy to replace lost brain cells is an experimental approach not sustained in the long term because the implanted cells are plagued by the same degeneration process that provoked the loss of function in the first place. The future will focus on adapting treatments to suit the unique biochemical composition and medical history of each individual patient. Personalized treatments can become very important in situations where particular patients do not respond to the conventional treatment. Stem cell technology has tremendous potential for applications in personalized treatments if it utilizes the patient's native cells for tissue repair. This circumvents immunosuppressant interference and the cohort of serious side effects and related problems. There also exists the possibility to help patients suffering from strokes, Alzheimer's disease, Parkinson's disease, epilepsy, stress, leukemia or depression. In this respect, nanoneuroscience will deal with the retrieval of stem cells, their differentiation into neural cells, the integration of regenerated cells in absorbing matrices inside the brain tissue, the regeneration of neurotransmitter secreting cells, and the regeneration of the cells expressing protective factors for supporting the glial cells and surrounding tissue.

Cancerous neuron cells behave differently when attaching to a surface matrix compared with healthy neurons. Deciphering the difference in membrane potentials and complex interfacial processes using biosensors platforms might lead to a deeper understanding of the cellular processes involved in malignity. Anti-cancerous drugs can be tailored first using *in vitro* investigations. Life-threatening brain surgery and consecutive tumor regrowth might be prevented using such targeted treatments. Negative side effects of chemotherapy might be also avoided.

7.8 Regenerative Techniques in Neuroscience

Neuroregeneration implies the restoration of neural functionality. Injury, disease and genetic factors can lead to permanent loss of function in peripheral nerves in the spine or brain. In the central nervous system, severed axons never regenerate and the lost neurons are never replaced, unless they are olfactory neurons. For so many years recovery and brain repair has been a major goal in neuroscience, although progress has been slow in coming. Attempts to use stem cells derived from the olfactory system of the injured person and inject them in the affected area of the spinal cord may one day yield fruitful results. At the brain level, accidents, stroke or diseases such as Parkinson's are even more difficult to treat. For almost three decades, efforts have been focused on replacing the dying neurons with embryonic tissue in order to regenerate the cells that use dopamine as a neurotransmitter. New hopes are related to the use of totipotent embryonic stem cells which somehow receive the right signal from the surrounding tissue to derive the correct type of cell. To overcome ethical debates, umbilical cord cells or stem cells available in each brain might be an alternative source for collection and use. However, the problems related to the harvesting, growing and delivering of these cells to the right areas of the brain and, more than that, to manipulate and control their proliferation, are real technical difficulties for the research community. Until regeneration can be naturally achieved by personalized techniques, generic treatments involving the miniaturization of biocompatible implantable devices for drug release and electrical stimulation of specific brain areas are temporary solutions to consider. It is evident that brain regeneration will constitute a major area of activity in neuroscience for decades to come.

7.9 Prosthetic Implants

In the next decades it is expected that sensory and motor impairments will be alleviated as a direct result of the increasing knowledge accumulated from modern brain imaging techniques. Correcting lost sensory and motor functions will be possible by implementing electrode arrays in the motor cortex designed to detect the intention of movement and translate it directly into the moving operation. This will enable paralyzed individuals to move and control their limbs. The identified regions associated with so-called 'higher' functions such as cognition, affect, memory and decision-making will lead to the development of direct brain–computer interfaces in the future. The first steps represented by successful cochlear implants continued with the implanted electrodes for deep brain stimulation for treating essential tremor, Parkinson's disease and depression. New technologies will improve the interface between the electrodes and the brain tissue and the performance of the electrodes. Implanted electrode arrays in the visual cortex will be further miniaturized and refined, making possible to enable at least the detection of light and dark, form and movement, and greatly improve the quality of life for blind people.

7.10 Dementia, Alzheimer's Disease and Reversing the Ageing Process in the Brain

Progress in medicine and higher standards of living have greatly increased the life expectancy of the general population, at least in the developed countries, without addressing the alarming issues concerning cognitive decline and the characteristic frailty of old age. The old theory that the brain inexorably loses neurons without being able to replace them does not appear to hold water anymore. Regeneration does occur in the flexible structure of the brain, especially in the hippocampus, in the areas associated with memory and learning. However, the process is slow and is overcome by the loss of plasticity and massive loss of grey matter in some older people. It is not clear why this is not a general phenomenon and how sometimes lucidity and sharpness of mind is still retained, even in the presence of degenerative processes.

There are millions of people worldwide affected by dementia. Alzheimer's disease is accepted as the main cause, with people over 85 in the major risk category. Only a very small percentage of the cases are familial, having genetic causes; the rest are random and probably highly influenced by previous lifestyle choices concerning exercise, nutrition and social activities, and keeping an enthusiastic interest in learning new things. Environmental risks, gender and the number of previous head injuries seem to have an influence too. Cellular research can use advanced *in vitro* techniques to study the accumulation of the amyloid plaque, the biochemical contribution of amyloid precursor proteins or genetic relevance for the neuron death which leads to the disintegration of the cognitive structures of the brain. Some preliminary attempts were actually presented in this book. A pharmacological approach involving neurotrophic factors and specific drugs seems a better choice than stem cell therapy, because neuron death is widespread throughout the brain and not localized to limited regions as in Parkinson's. Existing drugs have only a palliative effect and major efforts are targeting the degenerative processes, based on new advances in the understanding of complex neurological and immunological mechanisms. More effective drugs are expected to appear in the next decade, not only to diminish the symptoms, but to stop or reverse the decline, offering the joy of an independent, normal and fulfilled life to senior citizens.

7.11 The Human Brain Project and Computer Simulation

We so much want to understand our minds. The shadows and the light, how we process emotion and information, and what happens in normal and pathological conditions. The latest news regarding the concerted efforts to map brain activity (BAM) neuron by neuron, electric impulse by electric impulse, in a simultaneous unfolding of spatial and temporal patterns of brain firing processes, is taking the scientific community by storm. The boldness of such an endeavor is clearly exciting scientists. The Human Brain Project in Europe aims to create a computer simulation of the entire brain. Realistically, we know that computational modeling and analysis of every spike from every neuron will generate a flood of data impossible to handle even by the most powerful computer systems presently available. Computational techniques to handle such an amount of data have still to be developed. Of course it is advisable to start with limited numbers of neurons in networks grown *in vitro*, to be followed by small brains of worms, insects and small animals. As in the Human Genome Project, the initial awkward methods can be accelerated to previously unimaginable results. The Human Connectome Project, using functional MRI to track how different regions of the brain interact is a good start, even if it doesn't have the spatial and temporal resolution needed for a very detailed mapping. Invasive implantable electrodes, nanowires and non-invasive reporters like nanoparticles functionalized to trace the desired molecular reactions represent interesting developments, although at this time it is not known how they disrupt the normal functioning of neurons in the brain and so they surely cannot be applied to humans at this stage.

Despite all the projected effort the all-encompassing problem remains, that is, what is the significance if we can see all neurons firing in real time? Will we find the key to decoding the brain? Such mapping techniques are meaningless if we are not able to relate electrical activity to behaviour. Will we be able to finally connect the brain with the mind, cognition and consciousness? What if, as with the Human Genome Project, we end with more questions than answers? The outcome is difficult to predict. The applications in artificial intelligence, brain–computer interfaces, and understanding of normal and diseased brain might be invaluable and extremely helpful, leading to better treatments of mental disease and thus alleviating the problems confronting the present healthcare system and society in general.

7.12 Conservative Perspectives: A Final Comment

The conventional strategy in modern neurobiology is to examine, using state-of-the-art techniques such as functional MRI, neurons or subcircuits of interconnected neurons, as well as their feedback and inhibitory mechanisms that would explain brain activities, including thinking, emotions and maybe even intuition, creativity and insight. The current research plan is to search for neural correlates based on the assumption that the rate and manner of firing correlate with conscious experiences. Today, scientists are working assiduously to discover the neural code with the concept in mind that the brain is just a 'meat computer' and all that is needed is the software code to understand how it is functioning. This raises extremely difficult questions such as how is the mean firing rate associated with experience or is the chemical behavior of neurotransmitters the reason for our rich interior life? The problem is that computed axial tomography (CAT) scans, positron emission tomography (PET) scans, functional MRI and related technologies have not yet answered the fundamental paradoxes in neuroscience such as the problem of how diverse perceptions occurring in distant parts of the brain are bound together into a single conscious experience.

This book was created with the intention to contribute to a better understanding of the huge potential that new technologies represent for advancing research in neuroscience. We still do not know so many things about the marvelous machinery that is the human brain. The relationship between brain and the mind is still such a grand mystery because sentiments, insights, feelings, thoughts and creativity are nowhere to be found in the electrochemical mush inside the brain tissue. We see subatomic particles appearing through waves of potentiality from an invisible nothingness and we still wonder about what the ultimate reality could be. In an astonishingly fine-tuned universe that appeared miraculously in about 10^{-43} seconds and in which physical constants are so ingeniously premeditated that a change of one part in a billion would have dispersed the whole cosmos, we still wonder if life was an accident. If the universe waited patiently 13.7 billion years for us to arrive and appreciate its beauty, we ask ourselves if our minds are random processes emerging from the molecular activity of the brain. The alternative to emptiness and oblivion is the sense of awe and wonder that Einstein confessed to when witnessing the complex intricacies and mysteries of the quantum reality. Curiosity and enthusiasm can lead us all to many more great scientific discoveries.

Subject Index

Locators in **bold** refer to figures and tables

acetylcholine 57, **58**
acetylcholinesterase inhibitors 123
acoustic wave devices
    detection 157–60
    transduction strategies 19–27, **21**, **22**, **24**, **26**, **27**, 33–4
action potentials 58–60, **59**, 88–90
activation, covalent binding 10, **10**
active pixel-sensor based microelectrode arrays (APS-MEA) 99, 102
adenosine triphosphate (ATP) 170, 175
adhesion, cell 60–5, 68–9, **70**, **71**
adjunct chemicals 2, 17, 39–40
    comparison of AW/SPR 33–4
    see also Ru complexes
adsorption, direct/linker 6–7
adsorption, unwanted components 2–3
afferent neurons 57
AFM (atomic force microscopy) 40, 78
age-related macular degeneration 186
aging, future scenarios to counteract 200
agonist versus antagonist interactions 3
alpha brain rhythms 164–5
Alzheimer's disease 122–4, 149–50, 183, 200

aminopropyltriethoxysilane (APTES) 10, 152
amperometry **9**, **16**, 16–17, **17**, 152
amyloid β-peptides (Aβ) 200
    impedance-based techniques 149–50, 153
    in vitro recording techniques 122–4
analyte 1
anatomy, neurons 55–7, **56**
animal experiments, avoiding 112, 116, 118–19, 124
antagonist versus agonist interactions 3
aortic smooth muscle cells (ASMC) 63, 78
APS-MEA (active pixel-sensor based microelectrode arrays) 99, 102
APTES (aminopropyltriethoxysilane) 10, 152
arginine **9**
Aricept 123, 197
ASMC (aortic smooth muscle cells) 63, 78
aspartic acid **9**
assembled monolayer chemistry 10–12, **11**
astrocytes 78–9, 174
a-syn (α-synuclein) 150, 153
atomic force microscopy (AFM) 40, 78

ATP (adenosine triphosphate) 170, 175
attachment chemistry, comparison of AW/SPR 33–4
automation, biosensors 2
avidin 7

Bacillus subtilis 39
BAM (brain activity mapping) 201
bare substrates, cell-substrate surface interactions 61–4
Beer–Lambert absorption photometry 29
Bell, Alexander Graham 39
betaferon 157–8, **158**, 168, **168**
bioactive coatings 190
biochips 147, **149**
   see also biomimetic interfaces
biocompatibility
   biomimetic interfaces 172, 181, **182**, 190
   cell-substrate surface interactions 61, 62, 65, 72, 80–2
   pacemakers 173
biomarkers 195
biomimetic interfaces,
   brain-electrode 172–5
   deep brain stimulation 175–80, **176**, **177**, **178**, **179**, 190
   electronic detection techniques **93**, 93–5
   future scenarios 189–91
   mobile scalp electrode system **184**
   motor cortex prostheses 180–3, **182**, **183**
   replacement of damaged brain components 183–6, **184**, **185**
   retinal prosthetic interfaces 186–9, **189**, 190
   see also cell-substrate surface interactions
biomimetic peptides 52
biomolecular binding 153, **153**

biosensors 4
   architecture 1–2, **2**
   genesis/historical perspectives **4**, 4–5
   in vitro recording techniques 115–26, **120**, **121**, **126**
   probe attachment methods 5–13
   properties 2–4
bipolar neurons 57, **57**
birefringence changes, nerve activation 154–5
BIT-FETs (branched intracellular nanotube field effect transistors) **134**, 134–5
blindness, retinal prosthetics 186–9, **189**, 190, 200
blood cells, cell-substrate surface interactions 80–2
blood–brain barrier 195, 196
Botulinum toxin 101
bovine serum albumin (BSA) 39
brain activity mapping (BAM) 201
brain-electrode interface
   see biomimetic interfaces
brain imaging, future scenarios 198
brain implants
   see implants, medical
brain oscillations/resonance 160–5, **162**
branched intracellular nanotube field effect transistors (BIT-FETs) **134**, 134–5

calcium ions 145
calibration 4
cancer assays 17
cancer cell research, future scenarios 199
capacitance, electrode implantation 177
carbon nanotubes (CNT) 132, **138**, 139, 190
CCD (charge-coupled device) cameras 33, 130

Subject Index

cell culture analogues (CCAs) 116
cell–electrode interfaces
   see biomimetic interfaces
cell-substrate surface
   interactions 50–1, 60–1
     bare substrates 61–4, **63**
     basic energy equations **52**
     biocompatibility 61, 62, 65, 72, 80–2
     extracellular matrix proteins 65–72, **67**, **68**, **70**, **71**
     polypeptide coatings **64**, 64–5
     substrate morphology 72–7, **73**
     substrate rigidity/elasticity 53, 77–9, **79**
     substrate surface parameters 51–3
central nervous system (CNS) **55**, 55
charge-based interactions 51–2, **52**
charge-coupled device (CCD) cameras 33, 130
charge transfer interfaces
   see biomimetic interfaces
chemical recognition sites
   see probes
chemical sensors 1–2
cholinergic (acetylcholine production) 57, **58**
chromophores, organic 132–3
chronoamperometry 17
close-packed monolayers 10–12, **11**
CMOS (complementary metal oxide semiconductors) 89, 95–7, **96**, **97**, 99, **100**, 102
CNS (central nervous system) **55**, 55
CNT (carbon nanotubes) 132, **138**, 139, 190
coatings, bioactive 190
cochlear implants 173
cognitive defects, post-surgical 80
cognitive enhancement, future scenarios 197–8
collagen 66
coma patients, deep brain stimulation 175
composite response equations 3

computer–brain interface
   see biomimetic interfaces
conductometry **9**
cone-ingrowth electrodes 106–7
consciousness 142–3, 194–5, 201
   neural correlates 186, 202
   neural oscillations/resonance 164
contact guidance, neurons 61
corneal wounds, healing rates 145
coronary artery disease 80, see also stent technology
coulometry **9**
covalent binding, probe attachment methods 8–10, **9**, **10**
CRGDS oligopeptides 66–7, **68**
cultured in vitro recording
   see in vitro recording techniques
Curie, Pierre and Marie 20
cybernics 183
cysteine **9**
cytoskeleton, cell 53

deep brain stimulation (DBS)
   complications 179–80
   electrode implantation 175–9, **176**, **177**, **178**, **179**
   future scenarios 190
   Parkinson's disease 175, 178, **179**, 180, 200
definition, piezoelectricity 19
dementia 200
   see also Alzheimer's disease
depression
   deep brain stimulation 175
   future scenarios 200
DES (drug eluting stent) technology 81
detection, neurotransmitters 115
   see also electronic detection techniques; in vitro recording techniques
device-brain interface
   see biomimetic interfaces
diamond 132
dip casting 6

dipoles 52, **52**
dipstick assays 2, 16–17
direct adsorption, probe attachment methods 6–7
DMEM (Dulbecco's modified eagle medium) **157**, 166
DMSO (dimethyl sulfoxide) 152, 166
DNA 8, 18
dopamine 57, **58**
doping effects, graphene-based biosensors 136
double layer capacitance 151–3
drug discovery, microelectrode arrays 116–18, **121**
drug eluting stent (DES) technology 81
DTSSP (3,30-dithiobis succinimidyl propionate) 152
duplex formation, amperometry **17**, 17
dyes, adjunct 2

ECs (endothelial cells) 74–5, **81**, 81–2
ECM
  see extracellular matrix
EDC-NHS (ethyl(dimethylaminopropyl) carbodiimide–*N*-hydroxysuccinimide) **10**, 10, 67
EEG (electroencephalography) 107, 164
efferent neurons 57
elasticity, cell-substrate surface interactions 53, 77–9, **79**
electric cell–substrate impedance sensing system (ECIS) 146, **147**
  see also impedance
electrical potential, action potentials 58–60, **59**
electrochemical impedance spectroscopy (EIS) 17–18, **18**, 152, 187
  see also impedance

electrochemical systems
  adjunct technology 40
  cell-substrate surface interactions 63
  electrode interface reactions 173–4
  neuron signaling 57
  pretreatment, stainless steel 62
  transduction strategies **9**, 13–19, **14–19**
electrode implantation
  deep brain stimulation 175–9, **176, 177, 178, 179**
  microfabricated electrode arrays 181, **183**
  see also implants, medical
electrode-brain interface
  see biomimetic interfaces
electroencephalography (EEG) 107, 164
electromagnetic acoustic wave sensor (EMPAS) 22, **24**, 24–5, 27
electromagnetic radiation, optical devices **28–33**, 28–39, **35, 37, 38**
  see also vibrational field
electron beam lithography 38
electronic detection techniques 88–92
  see also in vitro recording techniques
electroretinograms (ERGs) 125, **126**
electrostatic chemisorptive forces 64
electrostatic gating effects 136
electrostatic interactions
  basic energy equations **52**
  cell-substrate surface interactions 51–2
  probe attachment methods 6
ELISA (enzyme-linked immunosorbent assay) analysis 169
EMPAS (electromagnetic acoustic wave sensor) 22, **24**, 24–5, 27
encapsulation, probe attachment methods 7–8
endothelial cells (ECs) 74–5, **81**, 81–2

engineered surfaces 60
    see also surface parameters/
        properties
enhancement, cognitive, future
    scenarios 197–8
entrapment, probe attachment
    methods 7–8, **8**
enzyme-linked immunosorbent assay
    (ELISA) analysis 169
epilepsy 124–5
epiretinal electrode arrays 187–8
equations
    composite response 3
    electrochemical transduction
        systems 14
    electrostatic interactions **52**
    glucose oxidase electrode
        system 5
    Nernst equation 16
    piezoelectricity 20
    surface plasmon resonance 31
ERGs (electroretinograms) 125, **126**
ethics
    brain imaging 198
    cognitive enhancement 198
    genetic advances 196–7
    regenerative medicine 199
    replacement of damaged brain
        components 183
ethyl(dimethylaminopropyl)
    carbodiimide–N-
    hydroxysuccinimide (EDC-NHS)
    10, 10, 67
eukaryotic cells 50–1, 53–4
    extracellular matrix 50, **53**,
        53–4, **54**
    membrane lipid composition **52**
    see also cell-substrate surface
        interactions
evanescent waves, biosensors 29, **30**
extracellular matrix (ECM) 50, **53**,
    53–4, **54**
    acoustic wave detection 159
    implants, medical 173
    proteins 65–72, **67**, **68**, **70**, **71**
    topography 61

extracellular recording techniques
    see in vitro recording techniques
extrinsic devices, optical fibers **29**, 29

fast Fourier transform (FFT) 161
felbamate 125
ferrocene iron p-arene complex 16, **16**
fiber-optics 28–30, **28–30**
fibroblast cells (FCs) 75–7
fibronectin 54, 62, 66
field-effect transistors (FETs) 14–15,
    15, 25
    biomimetic interface 190
    graphene 135
    nanotubes/nanowires **134**, 134–5
    in vitro recording techniques
        95–7, **96**
5-hydroxytryptamine (serotonin) 57,
    58, 118
flexibility, cell-substrate surface
    interactions 53, 77–9, **79**
flow-through detection 2, 4, 33
fluorescence-based biosensors 29, **30**
fluorescence resonant energy transfer
    (FRET) sensors 37
fluorescent imaging 130
fluorescent quantum dot
    nanocrystals **131**, 131–2
focal adhesion, cell 60–5, 68–9, **70**, **71**
forskolin 169
Fourier finite element methods 177–8
Fourier transform infrared
    spectrometer 34
FRET (fluorescence resonant energy
    transfer) sensors 37
functional groups, biosensor
    immobilization of proteins 9, **9**
functional magnetic resonance
    imaging (fMRI) 130, 198
future scenarios 194, 201–2
    biomimetic interfaces 189–91
    brain imaging 198
    cognitive enhancement 197–8
    dementia/aging process 200
    genetics 196–7
    Human Brain Project 201

future scenarios (*continued*)
   nanoneuromedicine 195
   neuropharmacology 196
   prosthetic implants 199–200
   quantum neurobiology 194–5
   regenerative medicine 199
   stem cells 198–9
   vibrational field 169–70

GABA (γ-aminobutyric acid) **58**, 101, 112, 118
galantamine 123
gamma brain rhythms 165
gated ion channels 57, 58–9, 145, **158**
genetics, future scenarios 196–7
glass, cell-substrate surface interactions 68
glial cells 61, 198–9
   in vitro recording 90, 91, 99, 108, 109, **110**, 115
glial scars 174
glucagon, acoustic wave detection 159–60
glucose assays
   amperometry **4**, 4–5, 16
   dipsticks 2, 16–17
glutamate **58**
glutamate receptor
   memory formation 198
   neural oscillations/resonance 161–3
glutamic acid **9**
gold
   cell-substrate surface interactions 63, 68–9, 72
   comparison of AW/SPR 34
   nanoparticles 132
   self-assembled monolayers **11**, 11
   surface plasmon resonance 40
graphene-based biosensors 135–7, **136**, **137**, **138**
GRGDY (Gly-Arg-Gly-Asp-Tyr), cell-substrate surface interactions 68
growth cones, neuronal 61, 89–90

Hartman interferometry 34, **35**, 36
healing rates, corneal wounds 145
HeLa (transformed epithelial) cells 62
Helmholtz–Gouy–Chapman model 152
Hermann–Maugin notation, piezoelectricity 20, **21**
hippocampus 184, **185**, 200
histidine **9**
Human Brain Project 201
Human Connectome Project 201
Human Genome Project 196
human umbilical artery endothelial cells (HUAECs) 62
Huntington's disease 196
hybrid assistive limbs 181
hydrogels, cell-substrate surface interactions 65, **67**, 67–8
hydrogen bonding, probe attachment methods 6
hydrophobic interactions
   cell-substrate surface interactions 51–2
   probe attachment methods 6–7

IDTs (interdigital transducers) 25
IDES (inter-digitated electrode structures) 147
IEEE (Institute of Electrical and Electronics Engineers) notation, piezoelectricity 20
IGFET (insulated-gate field effect transistor) **15**, 15–16
imaging, brain 130, 198
immunochemical insulated-gate field effect transistor interactions (IMMUNOFET) **15**, 16
IMPs (integral membrane proteins) 10
impedance
   implants, medical 174, 177, 187, 190
   neurodegenerative diseases 148–9

sensing 145–54, **146**, **147**, **148**, **151**, **153**
spectroscopy 17–18, **18**, 152, 187
tomography, miniaturized 146, 147
implants, medical
   biocompatibility 172
   cell-substrate surface interactions 61, 73
   see also biomimetic interfaces; electrode implantation
in vitro recording techniques
   biosensors for neuroscience applications **120**, **121**, **126**
   charge transfer interfaces **93**, 93–5
   cultured neurons/neuro-electronic interface 88–92, **91**
   field effect transistors as neurotransducers 95–7, **96**
   interface fabrication methodology 105–10, **106**, **108**, **109**, **110**
   microelectrode arrays 97–102, **100**, **103**–**5**, 110–12, **114**, 116–26
   microelectronic interfaces 102–5, 115–16
   microfluidics 107–9, **108**, **109**, 112–15, **113**, **114**
   neuron field potentials 87–8
   neuroscience applications 115–26
   protocol to obtain dissociated culture of neurons from embryonic cells **92**
indium tin oxide (ITO)
   biochip 147, **149**
   microelectrode arrays 98
   nanostructures 139
   self-assembled monolayers 11
   sensors 1–2

infection
   electrode implantation 179–80
   retinal prosthetics 187
inflammatory responses
   biomimetic interfaces 190
   cell-substrate surface interactions 80
   implants, medical 174
   retinal prosthetics 187
inorganic nanoparticles 133, *see also* nanosensors, brain
Institute of Electrical and Electronics Engineers (IEEE) notation, piezoelectricity 20
insulated-gate field effect transistor (IGFET) **15**, 15–16
integral membrane proteins (IMPs) 10
integrated cell biochip microsystem platform, electric impedance sensing 147, **149**, **151**
integrins **54**, 54, 69, **70**, **71**
interdigital transducers (IDTs) 25
inter-digitated electrode structures (IDES) 147
interdisciplinary teams 143, 175, 194
interface, brain-electrode, *see* biomimetic interfaces
interference reagents 69
interferometry 34–6, **35**
intermodal segments, electrode implantation 176, **177**
intrinsic structures, optical fibers 29
ion channels, voltage-gated 57, 58–9, 145, **158**
ion fluxes 172
ion selective electrodes (ISEs) 14, **14**
iridium oxide film 190
'island' formations 6
ITO *see* indium tin oxide

Kelvin physics 40
   see also scanning Kelvin nanoprobe

label-free detection 2, 17
labeling, see adjunct chemicals
lab-on-a-chip structures 61
laminins 66
Langmuir–Blodgett film technique 10, **11**
light-addressable potentiometric sensor technology (LAPS) **156**, 156–7
linear multielectrodes 102
linker adsorption, probe attachment methods 6–7
lipid composition, eukaryotic cells **52**
lithography, cell-substrate surface interactions 73
Love-wave sensors 26
lysine **9**

Mach–Zehnder interferometry 34, **35**, 36
magnetic acoustic resonator sensor (MARS) device 27
magnetic resonance imaging (MRI) 198
magnetic transcranial stimulation 164
magnetism, adjunct technology 40
magnetoencephalography (MEG) 130
mapping, brain activity (BAM) 201
MARS (magnetic acoustic resonator sensor) device 27
mass response 3
MEAs, see microelectrode arrays
medical implants, see implants, medical
memantine 197
membrane electrodes 14
membrane lipid composition, eukaryotic cells **52**
membrane systems, close-packed monolayers 10–12, **11**
memory 123–4, 197–8, 200
MEMS (microelectromechanical system) 39, 180

metal oxide semiconductor field-effect transistor (MOSFET) 135
Michelson interferometer 34
microelectrode arrays (MEAs)
　basic neuroscience research 122–6, **126**
　drug discovery applications 116–18
　microelectronic interfaces 102–5, **103**
　substance fingerprint database 121
　toxicological applications 118–22, **120**
　in vitro recording 89, 95, **96**, 97–102, **100**, **105**, 110–12, **114**, 115
microelectromechanical system (MEMS) 39, 180
microelectronic circuits, *in vitro* recording 93, 94
microelectronic interfaces, *in vitro* recording 102–5, 115–16
microfabricated electrode arrays 181, **183**, see also biomimetic interfaces
microfluidics 107–9, **108–10**, 112–15, **113**, **114**, 195
microphotodiode arrays 188–9
microprobes, neural 181, **182**
　see also biomimetic interfaces
microscale cell culture analogues (CCAs) 116
mind, see consciousness
miniaturization, biosensors 2, 146, 147
　see also nanosensors; nanotechnology
mitochondria 53
mobile scalp electrode system **184**
molecular imprinted polymers (MIP) 12–13, **13**
molecular level neuromedicine see nanoneuromedicine
monolayers, close-packed 10–12, **11**

MOSFET (metal oxide
    semiconductor field-effect
    transistor) 135
motor cortex prostheses 180–3, **182**,
    183
MRI (magnetic resonance
    imaging) 198
multidisciplinary teams 143,
    175, 194
multiple sclerosis 157, 196
multipolar neurons **57**, 57
multi-walled carbon nanotubes 138
myelin attachment segments
    (MYSA) 176, **177**

nanoneuromedicine, future
    scenarios 195
nanoneuroscience 198–9
nanoparticles **131**, 131–2
nanoribbons 139, **140**
nanosensors, brain 130–1
    graphene-based biosensors
        135–7, **136**, **137**, **138**
    nanoparticles **131**, 131–2
    nanotubes/nanowires 132–5,
        133, **134**
    neuroscience
        applications 137–40, **138**
nanotechnology
    cell-substrate surface
        interactions **73**, 74–6
    electrochemical transduction
        systems 18–19, **19**
nanotoxicology 132, 139
nanotubes, see nanowires
nanowire field-effect transistors
    (NW-FETs) arrays 134
nanowires 132–5, **133**, **134**
    biomimetic interfaces
        189–90
    future scenarios 201
neointimal hyperplasia, stent
    technology 80–1
Nernst equation 16
neural correlates of
    consciousness 186, 202

neural networks
    microelectrode arrays
        104, **105**
    oscillations/resonance, neural
        160–5, **162**
    self-assembled 90,
        109–10, **111**
    spatio-temporal patterns 89,
        104, 108, 116, 122
    in vitro recording 87–8, 89–91,
        91, **92**, 101, 112
neural oscillations/resonance 160–5,
    162, 170
neural prostheses, see prostheses
neural regeneration, see regenerative
    medicine
neuroactive compounds, in vitro
    recording 100–1
neurodegenerative diseases
    future scenarios 196
    impedance-based
        techniques 148–9
    see also specific diseases
neuro-electronic interface, electronic
    detection techniques 88–95, **91**, **92**,
    93
    see also biomimetic interfaces
neurons 54–5
    action potentials 58–60, **59**
    anatomy/types 55–7, **56**
    cell-substrate surface
        interactions 61, 63
    field potentials 87–8
    implants, medical 173–4
    synaptic junction **56**
    tissue engineering 60
neuropharmacology, future
    scenarios 196
neuroplasticity 60
neuroregeneration, see regenerative
    medicine
neuroscience applications
    nanosensors, brain
        137–40, **138**
    in vitro recording
        techniques 115–26

neurotoxins 101, 112
    impedance-based
        techniques 150
    in vitro recording
        techniques 116, 118–22, 123
    see also toxicology
neurotransmitters **58**, 115
    see also specific types
neutravidin 7
NMDA (*N*-methyl-D-aspartate) 101, 115, 117, 123, 161–3
nodes of Ranvier 59, 94, 107, 176, **177**
norepinephrine **58**

oligonucleotides 9, 18
oligopeptides 66–7, **68**
optical devices **28–33**, 28–39, **35**, **37**, **38**
optical sensing platforms 154–7
optrodes 29, *see also* fiber-optics
organelles, cell 53
organic chromophores 132–3
organometallic complexes of Ru 2, 17, **17**, 40
oscillating neural networks 160–5, **162**, 170
oxidative stress, cell 133, 139, 174

pacemaker neurons 161–3
pacemakers 173, 176, 180
painting 6
paranodal main segments 176, **177**
Parkinson's disease
    deep brain stimulation 175, 178, **179**, 180, 200
    future scenarios 200
pattern formation, neuronal networks 87–8
PCBs (printed circuit boards) 150, **151**
PCL (poly(caprolactone) 75, 77
PDMS (poly(dimethylsiloxane) 61–2, 75, 107, 109, **111**, 113
PEG (polyethylene glycol) based hydrogels 65, **91**

peptides, cell-substrate surface interactions 52, 54, 65–72, **67**, **70**, **71**
peripheral nervous system 118
PGA (poly(glycolic acid)) fibers 75
pharmacogenetics, future scenarios 197
phenobarbital 125
photoacoustic spectroscopy 39
photo-lithography 38
piezoelectric physics 19–26, **21**, **22**, **24**, 139, **140**
planar microelectrode arrays (PEAs) 99, **104**, **106**, 106–7, **108**
plasticity 89–90, 92, 96, **97**, 130
plate mode devices 26
platelets 80–2
PLL (poly-L-lysine) **64**, 64–5
PMMA (poly(methyl methacrylate)) 75–7
poly(caprolactone) (PCL) 75, 77
poly(dimethylsiloxane), *see* PDMS
polyethylene glycol (PEG) 65, **91**
poly(glycolic acid) (PGA) fibers 75
poly(lactic/glycolic) acid (PLGA) 75, 77
poly-L-lysine (PLL) **64**, 64–5
polymerization 7–8
polymers 61–3, 66, 76, 78, 190
    see also specific examples
poly(methyl methacrylate) (PMMA) 75–7
polypeptide coatings **64**, 64–5
polysaccharides 52
post-surgical cognitive defects 80
potentiometry **9**, 13–16, **14**, **15**
power spectral density (PSD) measures 125
precision 3
pregnancy test dipsticks 2
printed circuit boards (PCBs) 150, **151**
probes 1, 2, 3
    attachment strategies 5–13, **8–13**

prostheses, neural 172–3
    cell-substrate surface interactions 60, 65
    future scenarios 199–200
    motor cortex prostheses 180–3
    nanostructures as scaffolds 137–9, **138**
    see also biomimetic interfaces; retinal prosthetics
protein A 7
protein aggregation/misfolding 149–50
proteins 9, **9**, 10
proteomics, future scenarios 197
PSD (power spectral density) measures 125
psychopharmaceuticals, future scenarios 197
pulsed amperometry 152

quantum biology 170, 194–5, 202
quantum dots (QDs) 36–7, **37**, **38**, **131**, 131–2
quartz crystal microbalance (QCM) 23
quartz electric balance 20

radiofrequency (RF) currents 27
Raman spectroscopy 37–9
Ranvier, nodes of 59, 94, 107, 176, **177**
Rayleigh waves 19, 25, **26**
real time flow-through detection 2, 4, 33
receptors, *see* probes
redox properties, Ru complexes 17, **17**
refractive index units (RIUs) 32, 34
regeneration sieves **106**, 106–7
regenerative medicine 60
    future scenarios 196, 199
    interdisciplinary approaches 175
    nanostructures as scaffolds 137–9, **138**
    neural 172
    see also biomimetic interface

resonance, neural 160–5, **162**, 170
response times properties, biosensors 4
retinal prosthetics 186–9, **189**, 190, 200
retinal studies, microelectrode arrays 125–6, **126**
retinitis pigmentosa 186
RF (radiofrequency) currents 27
RGD peptides 52, 54, 68–72, **70**, **71**
rigidity/elasticity, substrate 53, 77–9, **79**
RIUs (refractive index units) 32, 34
rivastigmine 123, 197
RNA 8
Ru (ruthenium) complexes 2, 17, **17**, 40

SAM (self-assembled monolayers) **11**, 11–12, 69
SAW (surface acoustic Rayleigh wave) sensor 19, 25–6, **26**, 27
scanning Kelvin nanoprobe (SKN) 40, **165**, 165–9, **166**, **167**
scar tissue 174, 191
selective binding 1
selectivity/specificity 3
self-referencing 4
self-assembled monolayers (SAM) **11**, 11–12, 69
self-assembled neural networks 90, 109–10, **111**
semiconductor technology 95–7, 131, **131**, 133, *see also in vitro* recording techniques; nanosensors
senses, human 142–3
sensitivity 3, 6
    comparison of AW/SPR 34
sensor devices 1–2, *see also* biosensors
sensorgrams 32, **32**
separation, two-step 113, **114**
serotonin 57, **58**, 118
SERS (surface enhanced Raman spectroscopy) 38

shear-horizontal surface acoustic wave (SH-SAW) device 26, **27**
silanization adlayer chemistry 72
silicon
    cell-substrate surface interactions 68, 70, 72
    CMOS technology 95
    comparison of AW/SPR 34
    nanotubes/nanowires **133**, 133–5
    in vitro recording techniques **91**
    self-assembled monolayers 11
single-walled carbon nanotubes **138**
SKN (scanning Kelvin nanoprobe) 40, **165**, 165–9, **166**, **167**
Smart microfluidic systems 195
smooth muscle cells (SMCs) 63, 75, 78, 81
sodium–potassium pump 58–60, 158–9
sol-gel technology 8
spatial crowding, self-assembled monolayers 11, **12**
spatio-temporal patterns, neural networks 89, 104, 108, 116, 122
spin coating 6
SPR (surface plasmon resonance) 8, 30–4, **31–3**, 154–6, **155**
spraying 6
stainless steel
    cell-substrate surface interactions 62–3
    stent technology 81
stem cells
    future scenarios 198–9
    microelectrode arrays 101
    nanotubes/nanowires **133**
    repopulation or replacement of brain cells 97
stent technology 62, 74, 80–2, **81**
streptavidin 7
STW (surface transverse wave) sensors 26, **27**
SU8 biocompatible polymer 181

subretinal electrode/microphotodiode arrays 187–8
substrate morphology 72–7, **73**
substrate rigidity/elasticity 53, 77–9, **79**
surface acoustic Rayleigh wave (SAW) sensor 19, 25–6, **26**, 27
surface enhanced Raman spectroscopy (SERS) 38
surface-launched acoustic wave biosensors 25–6, **26**
surface modifications 6
surface morphology 53, 72–7, **73**, *see also* topography
surface parameters/properties 51–3, 60, 173
surface plasmon resonance (SPR) 8, 30–4, **31–3**, 154–6, **155**
surface transverse wave (STW) sensors 26, **27**
synaptic junction **56**, 56–7, 87

tagging, *see* adjunct chemicals
target functional groups, immobilization proteins 9, **9**
target/analyte 1
tau/phosphorylated tau 149, *see also* amyloid β-peptides
tetrodotoxin (TTX) 101
thiamine 10
thickness shear mode (TSM) devices 19
    acoustic wave detection **157**, 157–60, **158**
    cell-substrate surface interactions **63**, **64**
    piezoelectricity 22, 22–5, **24**
    scanning Kelvin nanoprobe 166, 169
thiols 11, **11**
three-dimensional (3D)
    lithography 95
    matrices, PEG hydrogels **91**
    microelectrode arrays 102, **103**
3T3s (transformed 3T3 fibroblasts) 62

thrombogenicity, cell-substrate surface interactions 80, 81
tin oxide, *see* indium tin oxide
tissue capacitance 177
tissue engineering 60
tissue–device interface, *see* biomimetic interfaces
topography
    cell-substrate surface interactions 60, 61, 63, 72–7, **73**
    implants, medical 173
    see also surface morphology
toxicology
    microelectrode arrays 118–22, **120**
    nanomaterials 132, 139
    see also neurotoxins
transducer–probe combinations 3
transducers 1
transduction strategies
    acoustic wave devices 19–27, **21, 22, 24, 26–7, 33–4**
    electrochemical systems **9**, 13–19, **14–19**
    optical devices **28–33**, 28–39, **35, 37, 38**
transformed 3T3 fibroblasts (3T3s) 62
transformed osteoblast-like MC3T3-E1 cells 62
trichlorosilanes 11, **12**
tryptophan **9**
TSM, *see* thickness shear mode devices
TTX (tetrodotoxin) 101

two-dimensional (2D) microelectrode arrays 102, **103**
two-photon excitation techniques 130
two-step separation, microfluidics 113, **114**
tyrosine **9**

unipolar neurons **57**, 57
urea **14**, 14

van der Waals forces 6
vibrational field 142–4
    acoustic wave detection 157–60
    electric impedance sensing 145–54, **146, 147, 148, 151, 153**
    future scenarios 169–70
    neural oscillations/resonance 160–5, **162**, 170
    optical sensing platforms 154–7
    scanning Kelvin nanoprobe **165**, 165–9, **166, 167**
    vibrating probe technology **144**, 144–5, **145**
vision restoration, retinal prosthetics 186–9, **189**, 190, 200
voltage gated ion channels 57, 58–9, 145, **158**
voltage-based functional imaging 130
voltammetry **9**

Young interferometry 34, **35**, 36
Young's modulus assessment 62

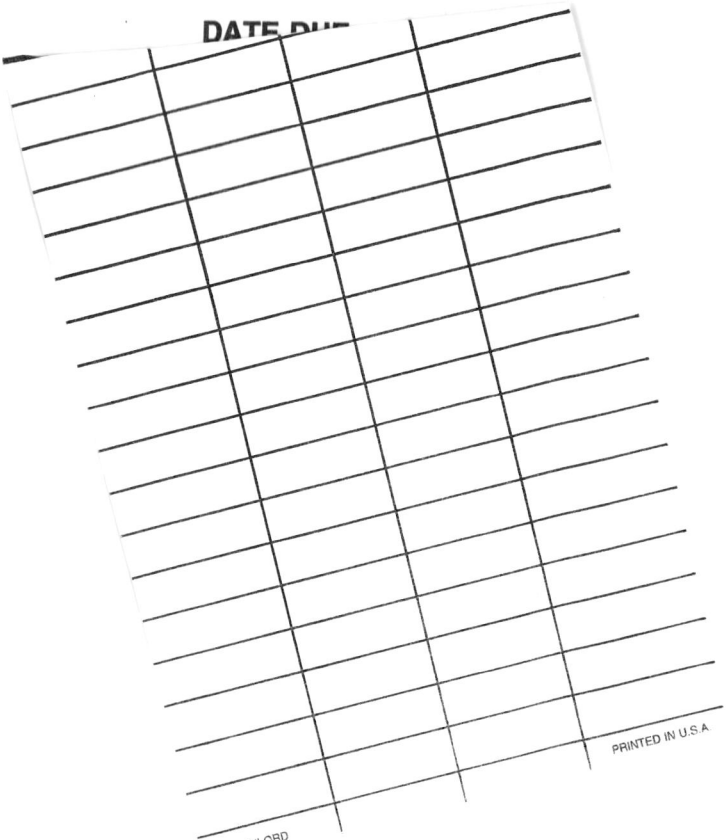